SECRETS
OF THE FATHER

The true story of a young man's search f
or his birth parents that lead him into
a scandalous cover-up in a small
New England town

By PAUL AUBIN
with SAMANTHA KELLER

Copyright © 2015 by Paul Aubin with Samantha Keller

SECRETS OF THE FATHER
The true story of a young man's search for his birth parents that lead him into a scandalous cover-up in a small New England town
by Paul Aubin with Samantha Keller

Printed in the United States of America.

ISBN 9781498453301

All rights reserved solely by the author. The author guarantees all contents are original and do not infringe upon the legal rights of any other person or work. No part of this book may be reproduced in any form without the permission of the author. The views expressed in this book are not necessarily those of the publisher.

Unless otherwise indicated, Scripture quotations taken from the Amplified Bible (AMP). Copyright © 1954, 1958, 1962, 1964, 1965, 1987 by The Lockman Foundation. Used by permission. All rights reserved.

Scripture quotations taken from the Contemporary English Version (CEV). Copyright © 1995 American Bible Society. Used by permission. All rights reserved.

Scripture quotations taken from the New King James Version (NKJV). Copyright © 1982 by Thomas Nelson, Inc. Used by permission. All rights reserved.

Scripture quotations taken from the New Living Translation (NLT). Copyright © 1996, 2004, 2007 by Tyndale House Foundation. Used by permission. All rights reserved.

www.xulonpress.com

Dedication

I would like to dedicate this book to all the adoptees, foster children, and those who grew up in homes never knowing their true identity or family history.

> *Ask and keep on asking and it will be given to you; seek and keep on seeking and you will find; knock and keep on knocking and the door will be opened to you.* Matthew 7:7 (AMP)

Acknowledgments

First I want to thank God for His inspiration and guidance throughout my journey.

My wife Margaret and the kids–They inspired me to take the manuscript I wrote a number of years ago, dust it off, and get it published so others could benefit from my story.

Amanda Aubin–My daughter who was a big part of the motivation to find my birth parents so that I could provide her with clarity on her own family history.

My stepchildren Hannah, Harrison, and Katie, as well as our foster daughter Molly, were all so supportive and encouraging.

Samantha Keller–Who is my good friend and a talented writer. She turned the search for my birth parents into a difficult to put down, wonderful, heart-felt story.

Barbara Sullivan–Who did a fabulous job of polishing up and editing the raw manuscript into the final version.

Florence and Frank Aubin–My adopted parents who are truly my real parents. They raised me, loved me, and cared for me as if I was their own flesh and blood.

Krystal Joy–A faithful friend who was there for me during every step of my journey, she inspired me to stay the course. Krystal was genuinely interested in the outcome around every corner.

Aunt Chris–My godmother in Massachusetts who helped me connect all the dots within her own family and who was always bold in uncovering the truth.

Karen Curtin–My adopted sister who was supportive and curious throughout my search for my birth parents. My search inspired her to go on her own journey to locate her birth parents in Ireland.

Nancy Sitterly–The private investigator from New England who did all the research and put the missing pieces of the puzzle together to help me discover my past.

Bibliography

Paul Aubin was given up for adoption by his birth parents when he was just five days old. His adopted family settled in Long Island, New York where Paul led a normal life in the 1960s and 1970s. He often wondered about the circumstances surrounding his adoption, but it wasn't until a health scare in the early 1990s that his interest in finding his birth parents and identity truly peaked. Paul went on a journey for several years to uncover not only his family medical history but also all the unanswered questions to the *why* behind his adoption. He never expected to run into so many roadblocks, detours, potholes, and land mines that were in place to protect the identity of his birth parents and throw him off track. This journey of faith and persistence changed Paul forever and gave him the answers to his past and the identity that he was seeking. He currently resides in Orange County, California where he is a pastor at a local church. He has heard God's calling to transition from a life focused on worldly success to a life of significance that will create a legacy of serving others. He and his beautiful wife, Margaret have a blended family that includes seven children and five grandchildren.

TABLE OF CONTENTS

Foreword... xiii
Introduction: A Story With No Beginning.................... xv
Prologue: Who Am I? xix

Chapter 1:   Odd Man Out 23
Chapter 2:   A Heart Crisis 31
Chapter 3:   The Struggle to Find 39
Chapter 4:   The Needle in the Haystack 47
Chapter 5:   Luck Runs Out............................... 51
Chapter 6:   Inspiration in the Dark 56
Chapter 7:   Snooping.................................... 68
Chapter 8:   The Gamble 75
Chapter 9:   The Problem With Lies 80
Chapter 10:   A Lack of Evidence 84
Chapter 11:   A Very Private Affair 93
Chapter 12:   The Cover Up................................102
Chapter 13:   Hidden in Plain Sight 115
Chapter 14:   Keys From the Past 125
Chapter 15:   A Beautiful Shell...........................125
Chapter 16:   Knowing a Good Man 140
Chapter 17:   Pomp and Circumstance 151
Chapter 18:   Familial Obstacles 172

Epilogue: Interview With Paul 177

Foreword
to come from Kenny Luck

Revelations are powerful.

They are like being in the presence of a volcanic explosion where bright lava, previously unseen, bursts onto the landscape of someone's life. If you are *not the subject* of a personal revelation there is a strange fascination with them. You can enjoy it from a safe emotional distance and be riveted by watching others be set on fire by new and dangerous truths. If you have ever been a *holder of a secret* there is the pressure of keeping it in – always pushing, bubbling, percolating, and pressing inexorably to the surface. If you are *the unfortunate subject* of an unwelcomed or spontaneous revelation – much like a volcanic event–there is no stopping its force once it released.

The truth comes out. A blast is heard. What was lurking below the surface just a second ago explodes outward. The once serene landscape of your reality is altered forever. Accompanying the fluorescent lava, black ash is rising high into the atmosphere and dominating the horizon. It's ashes fall everywhere. What once was is no longer. Everything will be forever different from now on. It is a spectacle. It is the purest shock and the most awful awe.

At ten years old Paul Aubin was caught in the blast zone of a most unexpected revelation and no aspect of his little soul would be spared. Imagine being told, *"You are not who you think you are."* And so begins the uneasy and private quest to answer the greatest existential question of the human experience: *"Who am I?"*

Secrets of the Father isn't a neat knock on the door by a long lost son. It's about a young boy, a small town, and a revelation behind the revelation that takes us all to places of our souls that need redeeming and rescuing. Some of us face parts of our past we want to hide. Others of us hunger inside to know why choices out of our control were made that changed so many things. Many of us go for decades unhealed and haunted. For all who have ever kept a secret, for all who have ever felt lost, and for all who simply want to see *someone* fight for and discover the truth – here is your man.

Kenny Luck

Introduction
A STORY WITH NO BEGINNING

As a kid, I often let my mind wander and dream about my biological parents. Of course, they were never ordinary imaginations—they were awesome and whimsical tales I told myself of why my parents had to give me up.

I envisioned my biological mother as a Nordic blonde with ice blue eyes. I visualized her running away at sixteen years old and joining the circus, where she fell violently in love with the brazen ringmaster who died unexpectedly in a ferocious lion attack. Heartbroken, pregnant, and now unable to climb the trapeze, I imagined my birth mother returning to her wealthy, socialite family, who had forced her to give me up. I pictured her then marrying a prominent businessman, who had no idea I existed. I could visualize her tear-stained face, as she dropped me off at the orphanage, a tiny bundle only a few days old. Furiously waving and weeping, she drove off in a black limousine, praying I would find a good home.

My fantasies allowed me to create an *alternative reality* where my parents had a good reason to give me up. I imagine most adopted kids play these mental games to cope with the feelings of uncertainty and abandonment that linger, in the back of the mind—that perplexing unsolved riddle of *"who am I and where did I come from?"*

Unlike many adoptees, I never obsessed over finding my biological parents. I enjoyed the love of a family who cherished me. Therefore, the yearning to find my true identity was possibly not as strong as those who grow up distinctly different or unsettled. As I entered adulthood, it dawned on me that there could be a compelling reason my parents chose to give me up, and it might not look like the pretty fabrications of youth. Many times private adoptions, where the birth parents' identities are concealed, imply a hush-hush backdoor type of affair. They can be shrouded with mystery and hidden because the truth is less than ideal.

For those reasons, I didn't seek to pursue my biological parents, until my heart literally forced me to. At the age of thirty-three, prompted by a cardiologist who diagnosed me with a rare medical condition, I began the search for my biological parents in order to obtain vital health information and my medical history. I knew the quest wouldn't be easy because it was a private adoption back in the late 1950s, in New England, with no public access to records and archaic filing systems. But I never dreamed my hunt would turn into the cat and mouse chase that unfolded. Roadblocks, detours, and cover-ups met me at every step. My private investigator was dumbfounded. She had never came up against a case where the identity of the birth father was intentionally enshrouded in secrecy. The more detours and roadblocks we encountered the more it peaked my curiosity and motivation to uncover the truth. I also had close family and friends who were as curious as I was about my birth parents. They encouraged me along the way, when I felt like I couldn't continue this emotionally exhausting search.

As I've shared my intriguing journey over the years with people, they have encouraged me to tell my story on a broader scale. They have told me that my experience could give hope and inspiration to adoptees desiring to find one or both of their parents or, for that matter, anyone in search of a birth parent. Knowing that so many people could benefit from the outcome of my search inspired me to continue.

The outcome of your search may not end in a triumphant, fairy tale embrace of long-lost endearments—or the scandalous cover-up I discovered. But I hope it will, at the very least, provide clarity and understanding to the questions all adoptees yearn to have answered. This book about my unconventional story is intended to provide real life examples of how I uncover the truth of my identity.

Additionally, for parents who have gone through the heartache of giving up their child for adoption, I hope this book gives an open and honest glimpse into the emotional highs and lows of an adopted child and their unique perspective. Please note I have also altered some of the names and places in this book to protect my family.

Join me as I delve into my own true story of the hidden secrets of a church and the desperation of a young man and woman caught in a web of fear and deceit; and the fierce desire of a barren couple to have children. As well as, my own spiritual pilgrimage to answer the age-old question of *who am I?*

Paul Aubin

PROLOGUE

WHO AM I?

We need to know where we came from. Knowing connects us, links us, and bonds us to something greater than we are. Knowing reminds us that we aren't floating on isolated ponds but on a grand river.

Max Lucado, *God's Story, Your Story*

It was the summer of 1963, just shy of my fifth birthday, when my parents loaded up the family car and we traveled down to the Babylon docks on the south side of Long Island, New York, not far from where we lived. My fingers brushed over the prized bouncy ball and jacks in my pocket, as I stared out the window at the lush trees and elaborate estates giving way to the wetlands and summer bungalows of the village. I wondered if my mom would let me climb down the muddy banks between the docks, if the tide was low, and skip rocks. Looking down at my best corduroy trousers, I wiggled in anticipation and tugged at my starched and buttoned up plaid shirt.

I grinned at my mom, as she turned and slicked back an errant strand of her frosted blond bouffant, hoping she would notice I was on my best behavior. But mother seemed distracted—her smile anxious and father's face was drawn and tight. Something big was brewing! Maybe we were getting a puppy or a new baby or, better yet, an ice cream cone from Friendly's after lunch? I could taste the ice cream in my mouth already, coffee—yum. But a puppy would be okay, too. A boy baby wouldn't be so bad, but a girl—yuck!

As my dad pulled our old 1952 Hillman to a stop and searched for an open parking spot, I pulled with all my might and cranked down the back window to get a better look. The docks smelled like fish, clams, and old wood mingled with brine. I breathed in the salty ocean air and exhaled little puffs of excitement. The second my dad turned off the ignition, I jumped out of the car and darted over to the edge of the street as my mom chided me to stay close. She grabbed my hand and the three of us wandered down the main drag towards the docks.

Babylon is an old East Coast waterfront village and the gateway to travel back and forth by ferry to the Barrier Islands. It's a blend of conservative and grand colonial homes with ramshackle fisheries, historic hotels, and quaint restaurants and shops. While beautiful, nothing in Babylon escapes the ravages of the sea and the unrelenting Atlantic storms. Its charm is precarious, predicated on the whims of the weather, thus making it all the more special to steal away to for an afternoon respite.

On a bench overlooking the harbor, my dad pulled me up onto his knee and my mom settled down beside him. She smoothed out her tailored, powder blue dress of imaginary wrinkles and

PROLOGUE

fiddled with her collar. Finally, she turned and grabbed my arm, her fingers lightly resting on me; she gave me a long searching look with watery eyes.

"Paul, you are a very special little boy," my mom said. "You are even more special than most children because you were chosen."

My dad's face, ever stoic, subtly nodded in agreement.

Confused, I looked back and forth between the two. This was not the puppy conversation I anticipated. My mom paused for a long, deep drawn out breath; my almost five-year-old mind could not grasp what was happening.

"What mommy?" I asked.

"Paul," she hoarsely whispered, "You are adopted. Your father and I, we adopted you."

Stunned, I shook my head back and forth, as my fingers tightened their grip on the bouncy ball in my pocket.

My mom leaned in closer; "Your birth parents couldn't keep you for reasons we don't fully understand. But, your dad and I, we love you very much and will support you and care for you like you were our own flesh and blood birth child."

I pulled the ball out of my pocket. Angry red marks bit into my skin where I had clenched the ball like a vice. I dropped it and slowly watched my favorite toy bounce down the docks, twisting and turning with each thud, then slowing in the cracks, until it reached the edge where my last glimpse of red slipped into the water.

The two of them looked at me. I stared back in a daze.

"Ummm, okay." I stuttered.

"Paul, are you okay?" said mom.

"Uh-huh," I gasped.

My mom grabbed my hand and helped me stand. Step by step we walked back towards the village. My feet felt like someone else's feet—someone I didn't know. Even Friendly's ice cream lost its usual appeal as the sticky cone dripped through my fingers onto the ground—a coffee cream trail of confusion. I wobbled and leaned against the woman I had thought of as my mom as questions pummeled my young mind like the waves crashing against the pylons of the docks.

Who are my birth parents? Did I look like them? Do I have brothers or sisters? Where do they live? Would I ever meet them? Why didn't my birth parents want me? Would the kids at school treat me differently—could they tell I was special?

Doubts consumed me; fear, loneliness, and an unfamiliar anxiety crept up within me. Would anything ever be the same, now that I knew the truth? Now that I knew I was adopted and this *knowing* interrupted everything?

Chapter 1

ODD MAN OUT

Who knows what true loneliness is — not the conventional word, but the naked terror? To the lonely themselves it wears a mask. The most miserable outcast hugs some memory or some illusion.

Joseph Conrad, *Under Western Eyes*

I slowly got used to the idea of my adopted status, but it was difficult. It was not unlike a child who grows up with a large birthmark. The stigma of being *adopted* set me apart from the other kids in a time and place where homogenous behavior was the unstated goal. Although being adopted isn't obvious on the outside, once my uptight East Coast neighborhood got a whiff of the juicy news, I became an outcast.

"Hey Paul, maybe you were hatched and the stork dropped you off at the wrong doorstep!" taunted the bullies on the block. To intimidate and torment me they even suggested that perhaps my parents were aliens from another planet who had dropped me off on earth temporarily. I was the kid everyone picked on, the butt of all the jokes, and an easy target for the bigger kids when no one was looking. It didn't help that I was a tiny slip of a thing and one of the smallest kids in the school. From the age of eight-years-old until puberty hit, unfortunately late, I was nicknamed *Peewee*.

My parents did their best to love and help me navigate through the awkwardness of growing up ostracized. But my pressing need to understand who I was and my questions regarding my birth remained unanswered. Because it was a private adoption, Frank and Florence Aubin, my adopted parents, knew almost nothing about my past and the circumstances surrounding my birth parents' decision to give me up. (Out of love and the greatest respect for my adopted mom, Flo, I will call her Mom throughout this book because that is what she has and always will be to me. I will also call my adopted father, Frank, Dad.)

On our way to visit relatives in Cambridge, Massachusetts, Mom and Dad would point out the old, crumbling brick orphanage off the parkway in Brooklyn, New York, where the stork apparently dropped me off. Every time we passed the orphanage, I said a prayer and thanked the Lord that my parents rescued me from that dump of a neighborhood. As bad as the bullies and snide comments were from the local kids, I knew my life could have been a lot worse.

We lived a comfortable life in a lovely home in a good neighborhood. My dad, Frank was an electrical engineer with a big company in Long Island, New York. Frank was a sharp guy—analytical and clever. He had a voracious appetite for learning and books; but he was highly introverted which made it difficult to connect with him on any emotional father and son level. Frank had little interest in throwing a ball, watching sports, or doing anything social. I was more like my mom, Florence, or as most people called her "Flo"—a social butterfly, athletic, and extremely vivacious. Although, I had a great deal of love and respect for my dad, it's my mom who captures my heart, draws me in, and understands me like no one else. Even though she couldn't physically give birth due to severe anemia, no birth mother could have cared more. There is an old adage that says, "Any woman can give birth to a child, but it takes a real woman to love, cherish, and raise that child." That's my mom Flo; she's a real woman with a mother's heart.

In 1967, when I was nine years old, my parents brought home a tiny lass from an orphanage in Ireland. Sweet Karen Marie was two years when she joined our family. The arrival of my baby sister was the culmination of my mother's dream of raising a little Irish girl. Florence's maiden name was Shanahan—as traditional as any O'Malley or Hennessey. Although she could have chosen to adopt anywhere, I imagine my mom longed for a deeper connection with her Irish roots and heritage. So, once I started elementary school, my mom and dad set out to adopt a baby girl directly from Ireland. It took almost two years for my adopted sister to arrive from an orphanage in Ireland, due to all the legalities and paperwork approvals needed for an overseas adoption.

I remember when she first arrived at the age of two—Karen had a very strong Irish brogue accent that was quite cute. Over time the Irish accent transitioned into a Long Island, New York accent. Although my sister and I had our fair share of fights and bickering, mainly due to the age and gender differences, Karen and I grew close as we matured. We were drawn together as adopted siblings and able to empathize with each other. Surprisingly, we looked alike even thought we were not related. She was also small in statute and empathized with my moniker of Peewee.

Just when I thought I couldn't take even one more day of being called Peewee, things changed drastically. I started to grow and grow, and then grew some more! In my sophomore year of high school I was barely four foot eleven inches and by the end of my junior year, I had shot up over fourteen inches, to six foot one. I was thrilled to be tall, but initially the sudden growth spurt didn't fit so well on my scrawny 135 lb. frame. Painfully skinny and scrawny, my arms and legs angled out like beanpoles. My new normal wasn't ideal—yet.

Not to be ungrateful for the burst of height, but if my birth father had been around I imagine he would have let me in on the family *height* secret. He might have said, "Paul, one day you will finally grow and it will be gangbusters growth. You won't be short forever." Ahhh…wishful thinking! This is a good example of one of the many mysteries we encounter as adoptees, not having any biological reference points as to what we may look like as we mature. Children growing up with their birth parents have a pretty good idea of what they are going to look like simply by seeing their parents, grandparents, and older siblings. They know approximately how tall they will be and how much they

...one of the many mysteries we encounter as adoptees, not having any biological reference points as to what we may look like as we mature.

will weigh. They get an idea of when that first gray hair will come in, or how soon they may go bald, and a variety of other traits that are passed down through the generations. In addition to the physical unknowns, as adoptees we struggle with having no medical history. Is there heart disease, cancer, diabetes, etc., in our family lineage? Whenever I visited doctors over the years and they asked for my family history I would have to leave that part of the form blank and explain later.

One day, my best buddy Glen Petrosino and I joined a gym in an attempt to put some much-needed meat on my bones. We worked out really hard and the *Peewee* moniker and all the teasing became my motivation for punishing weight-lifting sessions. At first I simply wanted to get fit, then I decided to work on adding some muscle, and then just for good measure—I decided to get really big.

As my newfound confidence settled in, I got into more than my fair share of fistfights. I even surprised (and scared) myself once I discovered how badly I could hurt people. During an attempted robbery at our local liquor store, I decided to take on the robber and teach him a lesson. One broken jaw, one dismantled nose, and a cracked skull later, they picked up the robber from the floor and took him away in an ambulance. The incident shook me up pretty badly and once I realized I could have killed the guy. I decided it was time to deal with all of the anger and resentment building over the years from being bullied, before my temper got the best of me.

Sports became my ally and an outlet for my angst over the years, especially as I moved into young adulthood. Although initially small, I was lightening fast and as I grew into my new larger frame, my ability to run and swerve from the big kids chasing me worked out to my benefit on the football field.

I spent every day after school playing sports in the street of my neighborhood. I played football, baseball, basketball, wiffle ball, bowling, tennis, and track and field. If the school offered a sport—Paul Aubin played it. By graduation, I was a four-time varsity letterman (not that I had anything to prove!).

The rest of my days were spent in the typical pursuits of an East Coast youth—Boy Scouts and camping trips, church, and enjoying the outdoors. My family attended Our Lady of Grace Catholic Church, where my mom was heavily involved. She taught catechism, wrote an article in the weekly church bulletin, sang in the choir, served communion, and organized fund-raising events. Boy, she was one busy lady! My mom was awarded the Saint Agnes Medal of Honor, given out annually to a diocese in New York, for an exceptional servant's heart over a lifetime. With over fifty-three years of service, she was the ultimate servant and as her friends and family always joke, "Everybody in church knows Flo Aubin." As a kid I was always known as *Flo's kid*. My dad was a trooper, always agreeing to join my mom in all the social engagements that she planned. Frank Aubin was an usher, so when I was old enough, I ushered too.

In the Aubin family, acquiring a strong work ethic was second only to God. And so by the age of twelve, I was out hitting the pavement, earning my keep and saving my pennies. I held some crazy jobs over the years, starting at 12 years of age with a newspaper route throughout my neighborhood. One day I was up to my ears in bubbles washing dishes in a restaurant and the next day I was handing out stinky shoes as a bowling alley attendant. I assembled trophy after trophy at that same bowling alley, mowed lawns, painted houses, and washed endless cars. I worked at the town pool in the summers, packed bloody meat in chilly freezers at the meat packing company, and was a waiter in several restaurants in town. No one could say Paul Aubin was lazy.

Working hard at school was another subject altogether. Academically I struggled. My classes failed to ignite any sort of interest; combined with poor study habits and boundless energy, my desire to excel in education stalled until college. That is when my motivation to hit the books finally kicked in.

In 1978, as I entered my twentieth year, opportunity knocked at the door. After attending a two-year community college in Long Island and continuing to pursue athletics, I received a scholarship to attend Ohio State University in Columbus, Ohio. This is where I buckled down and completed my bachelor's degree in the summer of 1981. I also managed to fit in a few moments of rip-roaring fun—okay—maybe a lot of moments of fun. My years at Ohio State were filled with playing sports, making life-long friends, and learning to make it on my own. They were some of the best years of my life.

Upon graduation, my roommate Billy Rodgers persuaded me to move all the way across the country to California. When I arrived in California, this East Coast boy thought the beaches of Orange County looked like an alien planet. But, I jumped right in and decided to experience the adventure for all it was worth and live a little while I was young. Billy and I found jobs at a health club selling memberships and embraced the beach lifestyle and laid-back California culture. I figured I could enjoy a few years on the West Coast and then head back to New York to get serious about a career and life. But, as usual, the Good Lord had other plans for me.

One day I signed up a breathtaking young woman and her friend for a gym membership. I have to admit she personified the Southern California dream girl. Two years later, I married the lovely Connie and three years later little Amanda was born. And with that one decision, to marry a California girl, all my plans of returning to New York fell by the wayside. Now I had a family—a real biological family. My baby girl Amanda was my first blood relative. And after all the years of subtle disconnects and nicks to my soul—my Amanda filled a large gap in my heart.

Chapter 2

A Heart Crisis

Asking is the beginning of receiving. Make sure you don't go to the ocean with a teaspoon. At least take a bucket so the kids won't laugh at you.

Jim Rohn, *Living With Abundance*
1990, Los Angeles.

All day long, out of the corner of my eye, I watched one person after the next avoid eye-contact with me and walk past my booth to wait in a long line for some new fangled test to determine their cholesterol levels. Jibs and jabs arose from the crowd mingling around the test-booth and soon it became an informal contest to see whose heart was the healthiest. Apparently this new discovery in the healthcare arena regarding the importance of having healthy cholesterol levels was gaining notoriety.

Now two years into my job as a pharmaceutical representative, I was extremely curious about the latest breakthroughs in heart research and all relevant medical terminology. I wanted to not only know; but also, intelligently speak the language of the cardiologists I sold pharmaceuticals too. This cholesterol phenomenon looked like the newest big thing and the testing booth was, by far, the most popular attraction at the medical conference. I decided to go over at lunch with a few of my colleagues and check out my own levels.

A few of the guys I worked with went ahead of me. Their numbers varied, but they were all in the one hundred and forty to one hundred and seventy range, which was perfectly normal. I smirked at my buddies as I stepped up to the plate. I was in prime shape—an athlete and gym rat—I had these guys beat on physical fitness all day long. Then the representative called out my number and my cocky grin turned upside down. My cholesterol reading was off the charts at three hundred and seven! What? It had to be a mistake. I thought this test had to be malfunctioning!

The company offering the testing stood by their assessment despite my resistance. They suggested I go down the aisle and around the corner to another booth that was doing a full blood workup. This test would require a fast for twelve hours to produce a more accurate reading. So, with my pride firmly tucked behind my tail, and now mildly concerned over my initial reading, I abstained from food that evening and had my blood drawn in the morning.

When the results came back in at—three hundred and ten—real panic set in. "What's wrong with me?" I asked the woman running the clinic.

She shook her head and examined my youthful and otherwise very healthy appearance. "Well, Paul, since you're only thirty-two and in good shape, I think you may have hyperlipidemia. It creates elevated levels of cholesterol due to a genetic abnormality. You probably inherited it. Do either of your parents have high cholesterol or cardiovascular disease?"

"I have no clue," I replied, "I'm adopted."

"Oh, well that makes it more complicated. Sorry," the woman responded shortly.

That old familiar ache of uncertainty that I'd spent thirty-two years trying to obliterate now pressed up into my throat once more. "I don't know the medical history of my family," I sighed.

The woman gave me a long piercing look and then shook her head once again at my sky high levels, "I suggest that given your extremely high reading you make an appointment with a cardiologist immediately."

I hopped off the table a bit defeated and walked back to my booth in a daze—so much for all those grueling workouts. I might as well have been a geriatric according to the test. What was going on with my heart? I needed answers and I needed them fast. My first thought was to tap into the friendships I had developed with some of the top cardiologists in Los Angeles.

One of the benefits of working in the pharmaceutical industry and specifically drugs geared towards the heart was open access to a multitude of the top cardiologists in Los Angeles. I decided to call Dr. O'Conner in Glendale, the premier heart doctor in my territory of Burbank and Glendale. Dr. O'Conner was in high demand and it was not easy to make an appointment. As a pharmaceutical rep, I had found it almost impossible to get an appointment but it was different now that I was a legitimate patient with a pressing need!

Dr. O'Conner ran a series of thorough blood tests, but once again my cholesterol remained at a shocking three hundred and ten. He asked me to go on a strict low cholesterol diet for ninety days and then come back for another reading. I took his advice seriously and wiped out the eggs, cheese, butter, and bacon from my diet. I worked out even harder, watched every bite, and was confident the next reading would be drastically different.

After ninety days of physical torture, my levels came back in at a whopping three hundred and nine. After all that work, I was still in the same boat. Dr. O'Conner, based on my reading, diagnosed me with familial hyperlipidemia, exactly what the clinician back at the cholesterol booth had suggested.

Knowing my story, which I shared at our initial consultation, and understanding my lack of clarity from a closed adoption, Dr. O'Conner was careful with his words as he approached me. "Paul, to gain insight into your health condition, I think it's time you began a search for your biological parents."

He glanced down at his records, "If either one of your parents had high cholesterol which led to heart disease and possibly death at a premature age, then I need to treat your condition much more aggressively. On the other hand, if they had high cholesterol and no further complications, then maybe cholesterol lowering medication and monitoring will suffice. For now, I'm putting you on a new drug called Lovastatin (Mevacor) in an effort to lower your numbers."

"Okay," I nodded, not having a clue where to begin.

I walked into the doctor's office with a heart condition and walked out on a mission to find my biological parents. I shook my head in bewilderment and thought, where do I start?

It's not like adoptee information is offered in the phone book. I was on the West Coast but my private adoption took place in the Northeast. It would be challenging, to say the least, but at the same time I felt strangely confident that this medical crisis was some sort of a sign from God to begin the search for my parents. At this time I was not the most religious guy, but religious enough to know God knew better than I did. So I closed my eyes and prayed, "God, I need a place to begin. I don't know where to start."

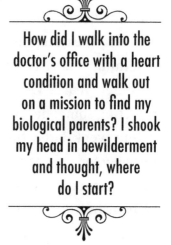

How did I walk into the doctor's office with a heart condition and walk out on a mission to find my biological parents? I shook my head in bewilderment and thought, where do I start?

I called my adoptive parents that night and told a fib. Okay, it might have been a lie, but it was a little white lie for a good reason. I wasn't ready to share my search for my biological parents with my adopted parents yet. It was still too tender and my emotions were raw. I told my mom I was traveling to England for work and I needed a passport, so they needed to dig up my birth certificate. They did and once it arrived in the mail, I ripped it open and pored over the document line by line. When I got to city of birth, I paused, dumbfounded. The birth certificate read *Place of Birth: Cambridge, Massachusetts*. All this time, I believed I was born in Brooklyn, New York. Confused, I immediately called my mom for clarification.

"Oh, Paul, that's right, you were actually born in Cambridge, Massachusetts, but all of your official adoption papers were processed in Brooklyn, New York. I'm sure that must have been confusing to you, so sorry, sweetheart."

I got off the phone and paced around the room, I couldn't believe what I was hearing. All these years I was told that I was born in Brooklyn, New York, only to find out at the age of 32 that I was actually from the same hometown as my parents, Cambridge, Mass. Why the deception and the casual, hey, no big deal attitude? Something wasn't right. Why would my mom deceive me for thirty-two years? Were my parents covering up something or protecting someone? It just didn't make any sense.

One week later, the company I worked for asked me to interview for a new position within the company—a promotion to a Managed Care Specialist. The interview was near the company headquarters in Overland Park, Kansas, at the Marriott. The interview took about an hour and then I had about forty-five minutes of downtime before the driver returned to take me back to the airport.

Feeling good about the interview, and a little restless from the adrenaline of putting my best foot forward, I decided to walk around the hotel and burn off some energy. When I entered the lobby, the marquee of companies holding conferences at the hotel caught my eye. I stopped dead in my tracks when I read the listing. The first company on the marquee was the International Adoptee Search Foundation. Huh? What are the chances? For the last week, I'd spent every night in agony thinking about my next step and fretting over where to begin. And now it seemed God had placed this information, like an amazing gift, into my lap. I have to be honest; this was probably the first significant event in my life, where I knew without a doubt, it wasn't just a coincidence. A shiver went up my spine and like a magnet my feet picked up and ran towards the room where the meeting was being held.

Panting, I walked up to the registration table where two nice ladies greeted me. "Go inside, dear. Check things out," they urged.

I opened the cavernous doors and shoved my shaking hands in my pockets. Glancing around, I realized the audience was filled with people just like me; they were all searching for their birth parents. They were just as apprehensive and exhilarated as I. They were also seeking a place to start and needing step-by-step guidance, and possibly some handholding to stave off emotional turmoil. I felt the turmoil rumbling in my heart. I slunk into a chair at the back of the room and listened attentively.

The guest speaker poignantly explained his story about searching for his biological parents. He was surprisingly funny and yet, brutally honest as he gave tips on the best practices to use to find biological parents. He also spoke about what to expect and not to expect in the search.

I noticed a table in the back of the room displaying books about adoption. I looked around the room at all the eager faces. And for the first time, I felt understood. They knew, in their own way, the ache I knew. We shared a similar story of uncertainty and it struck me that these faces asked themselves the very same questions I did! *Who are my parents and why did they give me up?*

I walked over to the book table as the speaker wrapped up, glancing down at my watch and wishing my remaining minutes would go slower. The lady behind the table grinned and asked, "Where were you born?"

"Cambridge, Massachusetts."

"That's wonderful," the woman replied. "I have a fantastic person to connect you with in the New England area. Her name is Nancy and she is very successful in assisting adoptees in their search to locate their birth parents. "In fact," she leaned in and whispered, "She is an adoptee too and has two adopted children of her own. She knows what you are going through and she can help."

I grabbed the pen and wrote down Nancy's contact information, and then bolted for my car and waiting driver. As I sprinted out of the hotel, I couldn't believe what had just occurred. It didn't seem real. On the way to the airport, I leaned back in the seat and held up the paper with Nancy's information, tears slipped out of the corners of my eyes and ran down onto my suit.

In that moment, it occurred to me maybe God really does care about me. Maybe He cares in more than just a life and death sort of way or more than measuring me on a scale of performing more good than bad deeds. I chewed on my thoughts, but deep down, my inner man was rocking and reeling. For a kid who had always felt different—like an outsider—this personal God directing my every step was something new and refreshing, both frightening and overwhelming. I didn't know what to think anymore, but I couldn't wait to call Nancy.

Chapter 3

THE STRUGGLE TO FIND

Genealogy becomes a mania, an obsessive struggle to penetrate the past and snatch meaning from an infinity of names. At some point the search becomes futile – there is nothing left to find... The only way to find out is to look at everything, because it is often when the searcher has gone far beyond the border of futility that he finds the object he never knew he was looking for.

Henry Wiencek, *The Hairstons: An American Family in Black and White*

Today, an adoptee searching for a birth parent doesn't ask, "Where do I start?" They simply open up their laptop and use Google to figure out how to begin. In the amount of time it takes to click a mouse, a host of resources will pop up. There are birth parent and adoptee registries, a host of birth parent investigators, online forums and support groups for those adoptees seeking information. The problem of this generation is not a lack of information, but an overabundance of data to wade through. In fact, it's hard to disappear these days and with the advent of social media and Radio-Frequency Identification (RFID) chips it's nearly impossible to hide anymore, or for that matter, to even protect or maintain a sense of privacy. Secrets in our modern world are scarce.

But this was not the case when I began my search for my birth parents. It was the early 1990s, a pre-Internet world without chat rooms, online networks or search engines. There were no Facebook support groups for adoptees, no Twitter or Instagram. You couldn't do an online shout-out for a missing person or have your *"I'm searching for my birth parent, help me!"* videos go viral on YouTube. I had to do the grunt footwork and research in an era before Starbucks and instant anything.

The problem of this generation is not a lack of information, but an overabundance of data to wade through.

The 2010 US Census Bureau cites two percent of children under the age of eighteen are adopted—about one and a half million. If you add in their biological parents, adoptive parents and siblings, it means one in eight Americans are directly touched by adoption. [1]According to surveys, the majority of these adoptees and birth parents have, at some point, actively searched for their biological parents or children separated by adoption.

---

[1] https://davethomasfoundation.org/wp-content/uploads/2012/10/ExecSummary_NatlFosterCareAdoptionAttitudesSurvey.pdf ."National Foster Care Adoption Attitudes Survey" Page 5; Dave Thomas Foundation.

Just like me, those affected by birth parent searching do so, for many different reasons; medical knowledge, the desire to know their birth parents as individuals, or because of a major life event that they want to share; such as, the death of an adoptive parent or the birth of a child. The most common reason; however, is genetic curiosity. Innate in humans is a driving desire to discover what their birth parents look like, what their talents are, and the kind of personality they have.

Innate in humans is a driving desire to discover what their birth parents look like, what their talents are, and the kind of personality they have.

When I first called Nancy, she gently led me through a series of questions and helped me to understand what this process might look like. The first objective of my adoption search was to discover the names of my birth parents. She also let me know that it would most likely be an emotional roller coaster full of mountain top highs and devastating lows. Nancy had a pleasant demeanor and seemed confident and ready to tackle this assignment. Based upon the information I had given her, Nancy recommended focusing the search on babies given up for adoption in Worcester, Massachusetts, on October 14, 1958. She also told me not to hold my breath. The search could take anywhere from several weeks to several months. In the meantime, she asked me to sign up for the National

...it would most likely be an emotional roller coaster full of mountain top highs and devastating lows.

Adoption Registry in case my biological parents or siblings had previously initiated a search for me.

Before I officially made the decision to hire Nancy as my investigator, I decided to meet her in person. I flew to New York and drove up to Connecticut to make the introduction. I brought my wife Connie along with my adoptee sister Karen. Karen was particularly interested in the birth parent search process. Nancy laid out her plans. She was meticulous and aggressive: clearly this was more than a job. It was her passion! Seeing her passion put me at ease. I decided to move forward with her services. I also decided that until I had more concrete information, I wasn't going to inform my adoptive parents about the search process. I wanted to avoid any undue emotional toll on them. This was tough enough for me to go through; I didn't see the point of involving them until closer to the end of the search process. I also needed to be very focused, without any distractions. Nancy was very knowledgeable of this process and explained the various adoption options to me.

There are two types of adoption—closed adoptions and open adoptions. A closed adoption is often referred to as "confidential" or "secret" and maintains the privacy of the biological birth parents, concealing their identity from both the adopting parents and the adoptee. This cuts down on any surprising disruptions for the birth parents later on in life. The records are tightly sealed and frequently the biological father is not recorded—even on the

In my opinion, closed adoptions do more harm than good. Adoptees should have the right to access their own medical history.

official birth certificate. Until recent years, closed adoptions were the more traditional and popular choices for Americans, peaking in the post-World War II Baby Boom Era. And while they remain as an option today, the practice of open adoption has gained popularity. In my opinion, closed adoptions do more harm than good. Adoptees should have the right to access their own medical history. It's no longer a cultural taboo to have a child out of wedlock and the social ostracizing towards unmarried women has lost its foothold.

In earlier days, before the advent of Internet technology, concealing birth records effectively discouraged the adoptee and the biological parents from searching for each other. If they desired to begin a search an investigator or private company was the only option. Since my adoption was done privately—as a closed adoption—there were no public records for Nancy to research or access. She was starting at the very beginning based upon the slim facts I had shared with her and the limited knowledge I had uncovered based on stories told by family members.

Nancy had to use a variety of sources to find the delicate information she sought. In my case, she had a contact at the Massachusetts State Department for Adoption. This individual oversaw the records department of children given up for adoption. With no Internet, Nancy had to investigate the hard way by searching for information in the local courthouses, libraries, and newspaper announcements that chronicle weddings and obituaries.

After several months of painful waiting on my part, Nancy telephoned me. "Paul, you might want to sit down. I want to share what I've discovered."

I found a chair and tried not to drop the phone.

"The name your birth parents gave you was Michael Patrick Fitzpatrick and you were born on October 14, 1958, in Cambridge, Massachusetts. Your birth mother's name is Barbara Stoll and your birth father is Walter O'Brien. If my records are still correct, they still live in Cambridge."

"Nancy, do you have their phone number?" I said trembling. "I want to contact them immediately."

"Paul, I know how excited and overwhelmed you are right now," Nancy said. "But I think you should wait a few days and let your nerves and emotions settle before you make maybe the most important call of your life."

"I think you should wait a few days and let your nerves and emotions settle before you make maybe the most important call of your life."

I knew Nancy held a Masters Degree in Psychology / Counseling, so although, my heart wanted to race ahead and pick up the phone, I agreed to heed her wise advice. Nancy proved invaluable as a mentor and adoptee counselor when it came to first contact with a birth parent. In my excitement, I didn't really play out the scenario of how a person might respond if I rang them up and said, "Hi Mom, I'm your long lost son." It might not go well for either of us.

I didn't really play the scenario out of how a person might respond if I rang them up and said, "Hi Mom, I'm your long lost son." It might not go well for me.

The truth was, this was a complicated and delicate situation, and my desperate desire for relationship might be perceived as more of a launched grenade than an attempt at restoration.

Nancy took the time to coach me through a mock phone call to build my confidence and prepare me. These were Nancy's tips:

- Don't say, "Hi Dad!" or "Hi Mom!" They will be shocked, stunned and in complete denial. While you have had the luxury of preparation, he or she will be getting the news out of the blue and will more than likely, deny it in an effort to stall and gain time to digest the news.
- This effort may take more than one call. You will have to build trust. It may take several calls before a birth parent agrees to a face-to-face meeting.
- Keep the conversation confidential! This may be a well-guarded secret. There was a reason you were given up for a private and closed adoption. The people and family in their life may not be aware—kids, husbands, and wives—all may not be privy to this information.
- Don't get trapped in unrealistic expectations of who your parents are—what they look like, who they are, and if they will embrace you.
- Be patient and sensitive to their response to you barging into their lives so many years later.

I developed an outline with specific topics I wanted to cover and tried to create a general flow for the call.

I prepared for this conversation for three solid days. While I was giddy with anticipation, I knew I had to gain control of my emotions. The extra days helped me to process and digest the idea of calling the woman who gave birth to me thirty-two years ago. In an attempt to take out some of the intensity, I tried to think of this conversation like a tough sales call. I developed an outline with specific topics I wanted to cover and tried to create a general flow for the call. I thought about the different angles the conversation could take and prepared different questions, responses, and potential objections on her part. I looked at what I wanted to achieve on this call and made sure I was more than ready.

Chapter 4

The Needle In The Haystack

The way to find a needle in a haystack is to sit down.

Beryl Markham, *West with the Night*

After three days of intensive preparation, research, and practice I felt like it was time to go for it and make the call to my birth mother. After taking a deep breath and a self pep talk, I dialed the number Nancy gave me.

"Hello," a woman answered the phone.

"Is this Barbara O'Brien, formerly Barbara Stoll?" I inquired.

"Yes, it is, how can I help you?"

"Hi Barbara" I said, "this is Paul Aubin and I'm calling you long distance from California. I've been doing some research on my family tree and the records show we might be related."

Barbara responded with warmth, "Oh interesting! How do you think we might be related?"

"Well, ma'am, did you ever give up a baby for adoption back in the late 1950s?" I said.

"Oh my God, it's you!" she blurted out, "I have finally found my son! I knew you would seek me out some day. I prayed and prayed for you."

Trying to contain my emotions I continued with my outline, "So, just to confirm, you had a baby boy on October 14, 1958, in Cambridge, Massachusetts, and gave him up for adoption shortly after he was born?"

Barbara replied hesitantly, "I had a baby boy on October 14, 1956 not 1958 that I gave up for adoption."

Confused and now doubting the year I was born, I answered, "You're kidding? Are you sure it was 1956 and not 1958?"

My mind whirled, is it possible that I'm two years older than I thought?

"Oh yes, there is no doubt about the year," Barbara insisted.

Barbara and I continued to talk for at least an hour. She mentioned, "I would love to come out to California to visit you." I could hear the longing in her voice for me to be her long lost son.

I wanted it to be her, but something held me back and the lump in my chest resisted getting too close because the facts just didn't match up.

I told Barbara, "I will call you back after I re-checked all the records."

Immediately, I picked up the phone and dialed Nancy in frustration. I explained my phone call and Nancy was stunned and said, "Paul, are you absolutely sure you were born in Cambridge? Do you have a copy of your birth certificate?"

"Yes," I replied. "I'll mail you a copy."

Several days later, after Nancy received the birth certificate, she called me back. "Paul, after carefully examining your birth certificate I noticed there is no state seal stamped on it. This isn't an original. Do you have a copy of the original?"

"No," I replied.

"Well, why don't you contact your parents and ask them if they have the original certificate," she suggested.

"Paul," she lowered her voice, "I think there is some sort of cover-up going on here and somebody doesn't want you to know where you were born and who your parents are."

And then my big hands, so great at catching a football, dropped the telephone on my foot. All I wanted was to find my parents—not embark in espionage.

This new effort to retrieve my birth certificate required a fib to my parents, but I figured it was a small price to pay for the truth. I told them the Massachusetts Immigration office rejected my birth certificate because it didn't have the state seal. My mom, Flo, was very cooperative and mailed it to me the next day.

When I ripped open the envelope and studied the document, immediately the place of birth caught my eye—Westfield not Cambridge, Massachusetts!

Why would my parents lie to me? I immediately dialed my mom and confronted her about the inconsistency.

She calmly replied that I had been born at the hospital in Westfield, but I lived in Cambridge, which is where the home for unwed mothers was located. She blamed the error on sketchy details provided by her best friend Chris, and Chris' brother, Father O'Reilly.

I hung up the phone in disbelief and called Nancy back to share this new revelation. Nancy was livid because she had spent months researching my past in the wrong city. In fact, her frustration and this new twist cost me big-time. I had to shell out more money to begin a whole new search in Westfield, Massachusetts.

Possibly the worst part of this debacle was having to contact Barbara O'Brien to inform her that I was not the baby boy she gave up for adoption back in October of 1956. She was, understandably, extremely disappointed, as was I.

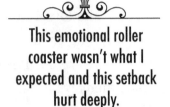

> This emotional roller coaster wasn't what I expected and this setback hurt deeply.

This emotional roller coaster wasn't what I expected and this setback hurt deeply. Barbara wished me the best in finding my birth parents and I encouraged her in the search for her son. Barbara seemed like such a kind woman and I hated to let her down, but now it was time to go back to the drawing board.

Chapter 5

LUCK RUNS OUT

If we will be quiet and ready enough, we shall find compensation in every disappointment.

Henry David Thoreau, *I to Myself: An Annotated Selection from the Journal of Henry D. Thoreau*

Months dragged by—Nancy did more research, in the correct city of Westfield this time around, and finally she called with news. Nancy had located my original birth certificate. It turned out there was another one besides the one my parents shared with me. If this was the first piece of the puzzle, now I was really confused.

Nancy uncovered the original document presented upon discharge from the hospital as a newborn at five-days-old. This document included the birth mother's name, the birth father's name and the original name my parents had given me.

Once again, Nancy asked, "Are you sitting down?" She then explained, "Your birth name is Michael Patrick Fitzpatrick (a moniker indicative of the New England area). Your birth mother's name is Margaret Fitzpatrick. Margaret Fitzpatrick is now Margaret Callaghan and she still resided in Cambridge."

Then Nancy paused, "Paul is it possible your adopted father is actually your birth father?"

"What? No! It's not even remotely possible," I retorted. "My adopted father looks nothing like me. He's about five foot eight and is of French Canadian descent. We don't remotely look alike."

"Well, this is very bizarre, because your adopted father's name is listed on your original birth certificate, but his middle name is incorrect." Nancy rifled through the papers, "It says Robert instead of Joseph. Again, I think there is something fishy going on here—some type of cover-up."

"It looks like your birth parents knew who they were giving you to—the Aubins—so they placed Francis Aubin's name in the line where it said birth father in an effort to conceal the identity of the birth father. However, they must have guessed on Frank's middle name. Its just more evidence to suggest a cover-up, Paul."

Nancy went on, "Your birth mother, Margaret is married to a fireman in Cambridge by the name of Michael and they have two daughters. Margaret works at St. Andrews Church as an administrative assistant. She is Swedish and Irish and both of her parents died many years ago."

"Paul, I think we should focus on contacting your birth mother and perhaps she can help lead us to your birth father."

When I could finally choke out a word, I agreed.

All of this information was too overwhelming to take in all at once. I had so many lingering questions; but now, I had more pieces to the puzzle. I was living in Cypress, California at the time and I hopped in my car and headed to the quaint little seaside town of Seal Beach. I needed to process and sort out the facts.

I parked my car and wandered down Main Street past the antique ships and pubs in a daze. Why would my birth father need to conceal his identity? Who was he? Oh, my gosh! My name is Michael Patrick Fitzpatrick and I have two sisters who probably don't know I exist. How old are they? Do they look like me? Will they want to meet me or have a relationship with me? Each answer brought more questions.

I stayed on the Seal Beach Pier that night for hours upon hours and thought more about the mystery man who was my birth father. Finally, chilled to the bone, I headed back to my car and returned home. I felt energized and optimistic that I was getting closer to the truth. I was wildly curious about my birth dad and thrilled to finally be able to contact my birth mother.

Only a few hours later, after a night of little sleep, Nancy called and began preparing me for the *big phone call* to my birth mother. She reiterated all the tips we had gone over earlier, before I had contacted Barbara. After much thought, I decided to use a similar tactic to my opening line with Barbara, "I'm doing some research on my family tree and I think we may be related."

I decided there was little need for me to shock and awe Margaret in the first thirty seconds of the phone call.

I decided there was little need for me to put Margaret into shock during the initial thirty seconds of the phone call. I wanted her to feel comfortable with me before I probed too deep. I put together an outline again, thought deeply about questions and possible responses and then once again, ran it past Nancy to review and change. Finally, I got the green light! It took some time to build up the courage to call, but eventually, I dialed the numbers.

A young woman, who sounded like she was in her twenties, answered the phone with a sweet and friendly New England accent and all of a sudden, my wall of no emotion crumbled. I hung up flustered. In all my preparation I had not considered my, more than likely, half sister answering the phone. This alone was exhilarating! When I finally recovered from my panic and excitement, I re-grouped and decided to try again in a few days.

The second time I called the same young woman answered, but this time I was ready. I asked to speak to Margaret and the young woman inquired who was calling. I said, "An old family friend," and she politely mentioned that "Peggy" (Peggy is short for Margaret) was available and called out for her mom to come to the phone.

"Hello," the woman answered.

With my heart in my throat I put together my words carefully, "Hi Peggy, my name is Paul Aubin and I'm calling you long distance from California. I'm doing research on my family tree for medical reasons and I think we might be related."

Peggy quietly spoke into the phone, "Ohhh?"

"Is your maiden name Fitzpatrick?" I asked.

"No, I'm not familiar with that name," she stated vehemently.

I tried again, "I'm so sorry, I must have the wrong woman—my apologies."

Just as I was about to hang up, Peggy started talking. She asked one question after another. For the next forty-five minutes she was very inquisitive.

"Why are you searching? Where were you raised? Do you have a family and children? Tell me about your career. Where do you live now?"

Peggy's curiosity was enough evidence for me. I instinctively knew she was my birth mom. I decided to take a risk, "Peggy, does October 14, 1958, mean anything to you?"

"Ummm, no, I, ahhh—should it?" Peggy stammered. The long stuttering hesitation pretty much confirmed to me that she was the one. She went on to say: "I know why you called me. There are two other Margaret Fitzpatrick's in this town and I'm always getting their mail and phone calls."

My stomach knotted at her inconsistent revelation. Initially, Peggy had said her maiden name was not Fitzpatrick.

"Peggy, I'm coming to town on business next week," I improvised on the spot. "Would you consider meeting with me?"

Peggy paused, "I—ummm, how about I check with the other two Margaret Fitzpatrick's to see what I can find out about before you arrive next week? You seem like a good young man with decent intentions, but you know this is highly sensitive and a lot of innocent people could get hurt. Call me when you arrive and we'll meet."

I thanked her, said goodbye and then hung up the phone. I was bursting with energy! I knew this was it! This was my biological mother and for the first time in thirty-five years I was talking to the woman who gave me birth. My blood felt like it was on fire and I was pumped with adrenaline and excitement. I couldn't wait to meet her!

Chapter 6

INSPIRATION IN THE DARK

Hope is tomorrow's veneer over today's disappointment.

Evan Esar, *American Humorist (1899–1995)*

A week passed, days of excruciating waiting, until I boarded the airplane for the East Coast. I arrived in Worcester eager to drive by Peggy's home and secretly check out where she lived. I also wanted to see if she was at home so that I could high tail it back to the hotel and give her a call.

I turned my rental car onto her street and held my breath. Peggy's home was at the end of a street at the top of a hill in a nice blue-collar neighborhood. It was a modest home that was well cared-for and comfortable. There were cars in the driveway, so I headed back to the Marriott Hotel and assembled my notes in preparation for our phone conversation. I was eager to confirm our meeting for the next day. After I checked in, I quickly unpacked and then picked up the phone, counted to ten, took a few deep breaths and dialed Peggy.

Peggy answered immediately and spoke in a hushed voice, "Paul, can I call you back later after everyone is asleep?"

"Of course," I replied.

It sounded to me like Peggy was hiding me from her family. If so, she was harboring some tremendously painful secrets from her husband and children. I couldn't imagine carrying around that burden for all these years. My heart hurt for this woman who lived in the shadows—never fully known and understood.

A few hours later, my phone rang. I picked it up and tried to play it cool.

"Paul, I've had some time to consider our discussion from last week. I don't think it's a good idea that we meet after all. This is too sensitive and I don't want anyone to get hurt."

I weakly responded with all the grace I could muster, "I understand Peggy and I respect your position. I would still like to stay in touch with you on this matter though, if that's okay?"

Peggy agreed and quietly hung up the phone.

I looked around the hotel room in dismay. I had traveled all the way across the country to meet with this woman, who was probably my birth mother, and she wouldn't even take the time to see me. A water bottle lying on the floor caught my eye and I kicked it into the bathroom, aimlessly looking to vent my frustration. Now what was I going to do with myself for the next week? I had planned on staying in Worcester for the rest of the week and then head across the Long Island Sound by ferry to visit my adopted parents and sister. This threw a big wrench in my plans and hurt more than I wanted to admit it.

As I tossed and turned in bed that night, I struggled to let go of my search when it seemed I was so close to the truth. By morning,

I'd decided that throwing in the towel at this point was a bad idea. I just needed to get creative and maybe a little sneaky.

I woke up bright and early and left the hotel around 6 am to swing by Peggy's house again, but this time with the intention of seeing her in person. I desperately wanted to know what my birth mother looked like. Maybe I could catch a glimpse of her walking out to her car or on her way to work. I also thought she might be more open to talking if I caught her at work instead of home. Her defenses and walls might be lower in a different environment.

I headed over to the bakery and grabbed a cup of coffee and a bagel to go, then drove to her street and parked at the end of it. I sat in the rental car holding up a local newspaper and trying to play it cool, while keeping an eye on the rearview mirror, just in case she pulled out of the driveway to go to work. I felt like the television private eye, Columbo, spying on a suspect, and I knew several of the neighbors were getting suspicious. I could feel the weight of their eyes on me as shades and blinds opened and concerned people peered through the cracks. I knew I didn't have much longer before someone called the police about a stranger parked in the neighborhood. And I fully expected a policeman to pull up behind me at any minute.

Finally, an automobile pulled out of the driveway and came down the hill approaching my car. I subtly glanced to my left and saw a woman who appeared to be in her late fifties driving the car. I quickly turned on the ignition and pulled out to follow her car at a safe distance. If I discovered where she worked then I could call information, get the number, and then call her at work to see if she might consider meeting me during her lunch break.

I followed the car for about ten miles, weaving in and out of traffic, trying not to lose her on the ancient and crooked streets of Worcester. The car pulled into an elementary school and stopped. Because Nancy had told me she worked at St. Andrews Church, I assumed the worst and figured I must have followed the wrong car because this definitely wasn't a church. So much for Columbo!

I raced back to Peggy's house to see if I could still catch her before she left for work, but by the time I arrived at her home both cars were gone from the driveway. Drats! Now what? It was 8:30 a.m. and Peggy probably wouldn't return until around 4:30 p.m. which left me with hours to fill.

Aunt Colleen from Worcester popped into my head and I drove back to the hotel and gave her a call. Aunt Colleen is my godmother and I called her *aunt* because of the closeness our families shared, not because she was truly my blood relative aunt. Her brother, Father O'Reilly was the Catholic priest responsible for bringing me from the home for unwed mothers in Westfield, to the Aubin's' home in Cambridge five days after my birth. I called Aunt Colleen for Father O'Reilly's current contact information. I thought he might be able to track down my birth dad and confirm my gut feeling about my birth mother Peggy. He had been in the thick of it from the very beginning. I didn't share with Aunt Colleen my reasons for contacting her brother, Father O'Reilly.

Aunt Chris was delighted to hear from me and gave me her brother's phone number. She encouraged me to give him a call and said he would love to see me. I called Father O'Reilly as soon as we hung up and he seemed genuinely excited to hear from me. I told him I was in town on business and asked if he was available to meet for lunch on Tuesday. Father O'Reilly said he was busy on that day and had a funeral on Wednesday, but he wondered if I could stick around until Thursday and catchup then. It had been a long time since we saw each other and he hoped I could make it. I agreed and hoped our meeting might reveal something I had overlooked along the way.

With plenty of time to kill, I gave Nancy (my private investigator) a call and updated her on my progress and my discussion with Father Ted. Nancy invited me to her home, just across the Connecticut border, for dinner with her and her daughter that evening. I accepted and then I headed over to the Naismith Memorial Basketball Hall of Fame in Springfield, Massachusetts, where I spent the afternoon happily perusing my favorite basketball heroes. Around 4:00 p.m. I headed back to Peggy's home and parked my car to wait for her to come home. Once again I was just hoping to get a glimpse of her, but after an hour of waiting and stalking there was no Peggy. I headed over to Nancy's for the evening meal.

At Nancy's house, it was finally time to relax and process the details of my trip over a glass of wine and some excellent barbeque. The conversation flowed nicely and her daughter was delightful. Together, we brainstormed why someone would cover up my birth father's name.

"Is it possible your adopted father is really your birth father?" Nancy inquired for the second time.

"There's no way," I replied. We look nothing alike and we share very few common interests."

"So, who is this mystery birth father?" Nancy's daughter shook her head in bewilderment.

After over an hour of brainstorming, all of a sudden Nancy's face turned ashen. "Paul, since Father O'Reilly was there from the beginning and he's the one who made arrangements to place the Aubins at the top of the list for newborn babies and he delivered you personally to them—well, is it possible he could be your birth father? Paul, do you know for sure that you were *really* in the home for unwed mothers, or was that a cover-up to protect the identity of Father O'Reilly? He knew you were going to the Aubins, so maybe he placed Frank Aubin's name on the birth certificate as the birth father. I bet he didn't know Frank's middle name."

Nancy jumped out of her chair with arms flailing and energy coursing through her body. "I'm going to drive up to his house in Springfield and knock on his front door pretending to be lost and looking for help with directions. I want to see if there's a resemblance!"

I thought the idea was crazy but I can't say I didn't pace the room nervously until she returned. About forty-five minutes later Nancy walked in the room with a wide smile like a Cheshire cat.

"Paul, he is the spitting image of you in about twenty years."

I gulped and my throat felt raw. I don't know if it was something I ate, a case of the flu, or the sudden shock that there was a good chance Father O'Reilly could be my birth father, but I suddenly fell ill. My temperature shot up and I ran to the restroom to be sick.

Nancy took me in for the evening and cared for me until the morning when I drove to the walk-in clinic to see a physician. I spent the next two days recuperating in the hotel and pondering the possibility of a Catholic priest as a birth father. Suddenly, it all started to make sense. The pieces of this jagged puzzle were falling into place.

Thursday, the morning of our meeting, finally arrived. I still felt wobbly from my illness, but with this new information I was more determined than ever to sit down with Father O'Reilly. On a whim, I decided to drive by his house to see where he lived and possibly catch a glimpse of him before we met. I tried to wrack my brain to remember what he looked like, but it had been at least ten years since our last in-person encounter. I found the address of the office building he gave me and nervously walked up as excitement churned in my belly. As I slowly strolled up, my eyes taking in every detail, I immediately spotted him. There were five men standing around him talking and yet after all this time, I knew exactly who he was.

Nancy was spot on with her assessment. Father O'Reilly was about my height, with bright blue eyes, bushy eyebrows, dimples, grey hair, and a big engaging smile. He looked like an older version of me and the resemblance was striking.

Father O'Reilly opened his arms and wrapped his large hands around me in a hug.

"Guys, this is my sister's godchild, Paul," he said with an endearing look.

I shook hands and politely introduced myself to the guys. Then we left his office, got in his car, and drove to a restaurant on the outskirts of town at the end of a country road. As we pulled up, I noticed how quaint the setting was—a typical New England style home now remodeled into a charming white cottage restaurant.

As we walked through the door, Father Ted motioned to the hostess to speak with her privately. He whispered to her, "Ma'am, can we please have a private table in the back of the dining room?"

Although he tried to keep his voice low, my ears were keen and I picked up the gist of his hushed tone. I thought it a bit odd, but it seemed Father Ted wanted to stay on the down low. So many secrets...

The conversation flowed easily. Father O'Reilly was an amiable guy and genuinely interested in me. We caught up on family, career, and friends. He wanted to know about my adopted parents since he had gone to school with them. He asked about my sister Karen, my wife and kids.

The waiter came to take our order. When Father O'Reilly requested the poached salmon with lemon, brown rice, margarine instead of butter, and fresh vegetables I almost fell off my chair.

I asked the waiter for the same meal and then lightly commented, "I'm impressed by your will power and commitment to eating well."

Father O'Reilly responded, "Well, I have an issue with high cholesterol so I've got to watch it pretty diligently."

"What's high?" I asked, holding my breath.

Father Ted made a sour face, "three hundred and seven."

I thought to myself, that is exactly what my cholesterol was on my last appointment!

I tried to disengage my former perception of him and look at him with fresh eyes. Did we share any of the same mannerisms or characteristics? I studied his bright blue eyes, nose, hairline, even his smile, as I tallied up a mental list of our similarities. I watched him gently brush back his hair from his ear in the way I do whenever I'm feeling anxious and chills went down my spine. During the conversation, Father Ted pressed his hand to his forehead to describe a stressful situation and again my inner alarm went off. It was eerily similar—even his voice sounded like mine, deep but soothing. His ministry years had served him well because his presence invited ease. He was personable, warm, and inviting and I couldn't help but relax around him because he made me feel like the most important person in the world during our time together.

"So, Paul, what are you doing in town?" Father O'Reilly inquired.

"Well," I paused and smiled, "I'm here on business, but since I'm all the way out here in Massachusetts, I thought I'd take advantage of it and do some research on my birth parents."

Father O'Reilly's rosy cheeks turned pale. His eyes blinked rapidly and although he tried to hide his initial reaction I knew I'd caught him off guard. It was the moment I was looking for, and by his reaction, I discerned the truth!

As Father O'Reilly coughed and tried to catch his breath, I continued, "My physician recently informed me that I have dangerously high cholesterol levels and recommended I find my birth parents medical history so he would know how to treat me. That way, if my birth dad died at an early age, he would obviously have to treat me more aggressively. But if my birth dad has high cholesterol and is still living well into his seventies, then my physician would place me on medication and monitor me on an annual basis."

Father O'Reilly slowly picked up his fork and took a bite. A thoughtful look crossed his face. "So, Paul, what is your plan to find your birth parents?"

"Well, Father O'Reilly, I was hoping that since you were there from the very beginning, you could shed some light on who my parents might be? I mean you did take me from the home of unwed mothers and deliver me to the Aubins, right?"

Father O'Reilly looked down at his salmon and closed his eyes for a moment in consideration. "Yes, Paul, you're right, I was there from the very beginning when your birth mom checked into the home for unwed mothers in Westfield. However, the vows that I took as a priest prevent me from sharing anything I may know about your birth parents. I hope you can understand?"

I nodded carefully, "Father O'Reilly. I completely respect your position but can you at the very least confirm that Margaret (Peggy) Fitzpatrick is my birth mother?"

"Paul, how did you come across that name?" Father O'Reilly looking and sounding a bit shell-shocked.

"Since I live across the country in Southern California, I'm physically unable to research my family history, so I hired a woman who lives here locally as a private investigator specializing in adoption searches."

The color in Father O'Reilly's face drained when I mentioned the private investigator.

Again I pushed, "Does that name Margaret Fitzpatrick sound familiar to you?"

Father O'Reilly reached for his water and took a long sip. He ran his fingers through his thick grey hair and then pressed his napkin to his forehead, dabbing at the beads of sweat forming on his brow. He didn't respond but sat quietly processing my words. Then finally he said, "You know Paul, come to think of it, her name does sound familiar—could be—but because of my vows I can't confirm anything. Sorry to disappoint you Paul."

I replied, "I'm just curious, have you talked to her or met her? Is she still in Cambridge?" Father O'Reilly leaned back in his chair and looked out the window. The sun darted behind a cloud and the room darkened briefly as he avoided my eyes, "Uh, Paul, did you ever meet your birth dad? Do you know who he is?"

I shook my head, "No, I have no idea who my birth dad is. I never met him. My birth certificate says Francis Aubin but we know that isn't true."

Father O'Reilly exhaled loudly, "Well Paul, I'll help you in any way I can."

I reached into my pocket and pulled out my wallet. "Father O'Reilly, here is my business card. It has a toll-free number on it (this is before cell phones) so you can contact me without having to pay long distance charges. If anything at all comes to mind regarding my family history, I would really appreciate you sharing it with me so that I can get to the bottom of my medical situation."

"Absolutely, Paul," Father O'Reilly affirmed. "I'll call you if anything comes to mind from over thirty years ago."

Father O'Reilly gave me a tentative grin, "Paul, have your adopted parents shared anything with you regarding your family history?"

There was sorrow in his question. I responded softly, "No, sir, all they have shared with me is that they hold you in the highest regard because you gave them something they could never have on their own—a newborn baby. They know that because you were childhood friends, you placed them at the top of the list of couples waiting for babies from the home for unwed mothers. You contacted them a couple of months prior to my being born in October 1958, and informed them they would have a baby around October 14. When I was five days old you delivered me to their home in Cambridge and then baptized me later that day. They know nothing about the birth parents or anything else about my family search. I decided not to inform them of this search because I don't want to upset them."

"Okay, Paul," said Father O'Reilly as he stood and grabbed the bill, leaving money for the meal, "I'll do what I can to help."

I thanked him, squeezed his hand and we headed back to his office where we said our goodbyes. Father O'Reilly stood and waved for a long time as I drove off. I gathered he wasn't thrilled about returning to the office after our intense conversation. I, on the other hand, was more convinced than ever that Nancy was right and Father O'Reilly was truly my birth father. But now I needed proof.

I drove back to the hotel in Cambridge in a daze. Trees and cars whizzed past as my mind ricocheted with questions and raw emotion. Amazingly, I pulled up to the hotel in one piece, with little recollection of how I got there.

I trudged up to my room, unlocked the door and collapsed on the bed. The red light on the hotel phone was blinking so I checked for messages and surprisingly heard Father O'Reilly's voice. He asked, "Would you come back to Springfield tomorrow so we can spend more time together?"

I was stunned. Would Father O'Reilly tell me the truth? I picked up the phone with trembling hands dialed and waited for my call to go through.

"Hello, this is Father O'Reilly."

"Hi, it's Paul and I would be delighted to meet with you tomorrow."

Chapter 7

SNOOPING

When you're curious, you find lots of interesting things to do.

Walt Disney

As long as I can remember, I've been curious about my birth—the hospital, the details, and all the things a parent naturally shares with their child that I missed out on. With all of the new information Nancy uncovered, along with our speculations about Father O'Reilly, I was more inquisitive than ever. I needed concrete confirmation of dates and times and records to piece this baffling mystery together.

What names were on the hospital record? What did I weigh? What time was I born? All of these thoughts pressed in on me after I got off the phone with Father O'Reilly. After thirty-five years of questions I wanted some answers and I was determined to get them.

Snooping

I decided to do some snooping on my own. According to Nancy, I was born at Noble General Hospital in Westfield, that wasn't too far away. I grabbed my keys and headed out the door to the rental car.

After crossing the threshold into the quaint New England town of Westfield, I checked my directions once more, slowed the car down and turned a corner. There stood Noble General Hospital, an antiquated brick building with ivy creeping up over the sides of the structure. It was straight out of an old movie and even though I'd grown up on the East Coast, after a decade in California, it still came as a shock to find such crumbling and outdated buildings still in use.

When I walked through the front doors, I tried to imagine what it looked like in the late 1950s, probably not that different. I found the information desk and asked the older gentleman manning the desk, "Where is the records department?"

The man directed me to the basement level for records but warned me to be careful and watch my steps because some of the stairs were a teeny bit rickety. I cautiously made my way down while trying to hold my breath, a musty and dank smell invading my nostrils. The lower I descended the darker it got. I stepped into a noiseless and empty corridor with no one in sight. Just as I began to question the old man's directions, I peered down the dimly lit hallway and saw a small sign that read "Medical Records".

Relieved, I opened the doors to the Medical Records department and there sat a woman perched on a chair like a frail old bird with a blue grey plume. She scooted up to the wobbly wooden desk with a big beaming smile and asked, "How may I assist you?" I got the impression that she hadn't had a visitor for quite sometime.

"Hello ma'am, I was born in this hospital and I need a copy of my records for medical reasons." I stated.

The woman seemed delighted to have a task. "Of course, young man, I'll just need your identification card and a five dollar copy fee."

"My name is Michael Patrick Fitzpatrick and I was born in 1958." I gave her the name Nancy had discovered, but it sounded strange to hear myself say my birth name out loud!

To my surprise she walked over to a file cabinet and stumbled through it. I thought, what no microfiche? No computer with stored records? Or for that matter, no old boxes full of musty paper from 1958? She simply pulled out a thin manila folder, walked over to the copy machine and proceeded to make copies.

I reached into my wallet and pulled out a crisp five-dollar bill. I could hear my heart beat hammering into my chest and it ricocheted behind my ears in a dull roar of yearning. I craved the truth with every ounce of my being and this frail little woman held what I desperately coveted without even knowing it.

"Hmmm," the woman noticed, "I wonder why your file is so small. It's only seven pages long?"

I quickly held out my five-dollar bill, hoping to distract and stall her from asking for my identification card. I tried to distract her by saying, "My family moved to Long Island, New York after my birth and I never returned to Noble General Hospital."

She slowly ambled over to me and laid down the copies on the desk in front of me.

It was my Indiana Jones moment. This was the Ark of the Covenant, my own personal Holy Grail, a priceless treasure greater than gold or diamonds—it was my knowing moment!

"Can I have your identification card, please?" she requested firmly.

I casually handed her my California Drivers License hoping she would give it a quick glance and not make the name difference connection.

It was my Indiana Jones moment. This was the Ark of the Covenant, my own personal Holy Grail, a priceless treasure greater than gold or diamonds—it was my knowing moment!

No such luck! This old gal was sharp. The woman glanced up at me with confusion. "This identification card doesn't match your name 'Michael Patrick Fitzpatrick'. Who are you?"

I gave her a slow lazy smile. I was hoping a little charm and flattery might get her to overlook the obvious. She wasn't impressed.

I dropped my voice and leaned in, "Please let me explain ma'am. At birth, my name was Michael Patrick Fitzpatrick, but then I was given up for adoption at five days old right after I left this very hospital. My adopted name is Paul Francis Aubin which is what you see on the identification card."

I looked into her eyes for a hint of empathy, but her scowl was deepening. I persisted, "You see, I have recently been diagnosed with a medical condition and my physician has strongly recommended that I obtain a copy of my birth records so he knows how to treat me."

The woman's chin jutted out in defiance like an old matriarchal turkey strutting around the farmyard. Her lips pursed, "I can't give you a copy of these medical records because there are policies in place that prevent me from doing so—even under your circumstances. I could get into a lot of trouble, so please hand them back to me!"

My hands clenched the papers even tighter. "Ma'am, with all due respect, I am Michael Patrick Fitzpatrick and these are *MY* medical records. I deserve to see my medical history. I'm afraid you won't be getting these medical records back from me."

I resolutely placed the five-dollar bill on the table, "Here is your five dollars for copy costs. I hope you understand my position. I'm so sorry. Thanks for your help and I hope you have a nice day."

The woman's mouth opened in distress—flabbergasted at my audacity to defy her orders.

With the records in hand, I dashed out the door, down the shadowy hall, bounded up the rickety stairs and lurched for the main door out of the hospital. I looked back as I sprinted for the rental car knowing security would be hot on my heels once the woman called for help. I was counting on her advanced age and the ancient security guard I saw on the way in to give me the advantage.

As I pulled out of the parking lot, I glanced in the rear-view mirror and noticed two security guards in their late seventies hovering near the front door of the hospital. They were looking for the perpetrator who dared to snatch the top-security medical files. Fortunately, their elderly pace didn't have me too concerned. Now they all had a story they could share of the young man racing out of the hospital with his medical records.

I picked up speed out of the parking lot and gunned my car for the highway. After putting some distance between myself and the hospital, I pulled over to review the records.

My hands trembled and the papers wobbled as my eyes eagerly scanned the document. I searched for Father O'Reilly's name to see if he was listed as my birth father—but he was not.

The signature for birth *father* read "Frances R. Aubin" just as Nancy indicated. And the "R" was the goldmine. This "R" was proof of a cover-up because my adopted dad's middle initial is "J" for Joseph.

Whoever filled out the birth certificate knew who my adopted father was going to be and guessed at his middle name. The birth father had handpicked the family who would raise me. This led me to believe with even more certainty that my birth father was Father O'Reilly. Father O'Reilly knew the Aubin's personally and understood I would be loved and cherished.

The birth mother read "Margaret Fitzpatrick" as her maiden name and Margaret Aubin as her newly married name. Her signature below substantiated it. This was very interesting because Margaret (Peggy) wasn't married and her last name definitely wasn't Aubin. This was another sign and con-

I had been waiting my entire life to answer the question of who is my birth mom?

firmation that there was a cover-up going on. And while Peggy might deny the truth, at least, I had bona fide evidence. This was a HUGE revelation and moment for me; I had been waiting my entire life to answer the question,"who is my birth mom?" The woman I had spoken to on the phone whom I suspected all along was my birth mother, was now confirmed. My initial instinct was to drive back to Worcester, knock on her door, and inform her of what I had uncovered. However, I eventually gained my composure and realized that I needed to be sensitive to her current family situation. Nobody knew her history of giving birth to me, but the birth certificate proved it. And, although, I'm not suggesting that anyone obtaining their birth records the way I did, it was one major missing piece to the puzzle regarding my past and identity.

I read on... I was born at 6:42 p.m. on October 14, 1958. I weighing in at 7 lbs. 4 oz. and was 19 inches long with no medical issues and a clean bill of health. The hospital released me five days after birth on October 19th—a surprising length of time. But I guess this was back in a day before managed care companies controlled costs.

It had the miniscule handprints and footprints of a tiny baby—but not any baby, they were MY handprints and footprints... Tears slipped out my eyes as I grasped the handprints and traced them with my finger.

I flipped through the final pages of the documents and the last one caught my eye. It had the miniscule handprints and footprints of a tiny baby—but not any baby, they were *my* handprints and footprints. It was another precious discovery—another random bit of information merging into my new paradigm. And a burst of quiet laughter erupted from deep within me. I sighed with relief, letting loose years of longing and ambiguity. Tears slid down my cheeks as I grasped the handprints and traced them with my finger. I thought about this feeling overcoming me—this wave of gratefulness arising from, well, *certainty*. It was at that moment I started to figure out who I really am.

Chapter 8

THE GAMBLE

Only those who will risk going too far can possibly find out how far one can go.

T.S. Eliot, Preface to *Transit of Venus: Poems* by Harry Crosby

Father O'Reilly suggested we meet at Friendly's Ice Cream Shop in Springfield, Massachusetts, close to his home. The irony of the ice cream shop was not lost on me. My adopted parents took me to Friendly's for an ice cream cone after they dropped the bomb on me about my adoption as a 5-year old. I wondered if today might bring a bit of redemption to my uneasy recollection of coffee ice cream cones.

I pulled into the quaint town of Springfield. The streets were lined with flag-waving colonials that were bedecked in red, white, and blue bunting as a tribute to the Memorial Day weekend. According to Father Ted, the holiday afforded him some extra time to spend with me. Whatever the reason, I was deeply grateful for another opportunity to get acquainted with the man who could be my birth father. I was also keyed up; eager to see if Father O'Reilly might disclose his past or keep under wraps the secret I believed he held close to his heart.

I parked the car and walked down the street to Friendly's, the sweet smell of waffle cones tempting my nose. Father O'Reilly was waiting with a wide smile and warm hug. He seemed genuinely pleased to see me. We grabbed a cup of coffee and a warm pastry, then sat down outside on a bench and chatted. I withheld my hospital escapade, despite its humor.

Father O'Reilly suggested we hop into the car and take a scenic drive through the historic Mohawk Trail that stretched from the Massachusetts-New York line to Millers Falls on the Connecticut River. I knew the highway meandered through the most picturesque areas of New England and I was delighted to go on a new adventure and spend the whole day with my probable birth dad.

The drive was leisurely and unhurried and Father O'Reilly peppered me with questions about my upbringing, interests, hobbies, career, family, and ambitions. Then we moved on to sports. I was delighted to dialogue about the Boston Red Sox, the New England Patriots, Boston Celtics, and all the New England sports teams. If Father Ted was my birth father then we both shared a love of sports, hallelujah!

Emboldened by the easygoing conversation, I asked him about his early days in school with my adopted parents, his call into the priesthood, his time in the Vietnam War, why he left the priesthood to pursue civilian life, his wife and kids, and finally, his medical history.

Father O'Reilly answered honestly, although at times he paused and picked his words carefully, but he never alluded to my adoption or who my birth parents might be. I had the sense it was on the tip of his tongue but he wasn't ready to disclose the past, yet. I tried to put myself in his shoes. What if I had a guarded secret for almost thirty-five years? And it was a big one! I was asking him to acknowledge that he had an illicit affair and impregnated a young woman of his church while he was a priest. And that he gave up the baby in a private adoption to friends who never knew he was the birth father. How hard would it be for me to bring that into the light? What if the collateral damage was too overwhelming to consider? What about his wife and kids who were in the dark? This was a scandal that would still be headline news if it were leaked to the press.

The tension of unspoken words hung in the air like the California June gloom, a misty haze of clouds and pressure systems veiling the sunshine. I tried to be patient and enjoy the moment without adding any more pressure to the battle I imagine was raging in his heart. I was confident, because of his integrity and the man I was getting to know that Father O'Reilly would come clean—eventually.

> ...without acknowledgement from Father Ted—my knowing meant nothing. I had a relational treasure in hand without the key to open it.

As the shadows began to deepen in the sky, we pulled back in front of Friendly's where my car was parked. It was difficult to say our goodbyes. We were so close to the truth and yet a million miles away from it all at the same time. I knew instinctively this was my birth dad and yet without acknowledgement from Father O'Reilly, my knowing meant nothing. I had a relational treasure in hand without the key to open it.

I said, "Father O'Reilly, thank you for a wonderful day. It meant so much to me," then I asked, "Do you still have my business card with my toll-free phone number? You can call me if you think of anything, any information at all about my birth parents."

I tried not to plead. I used my best non-emotive and stoic tone, but I'm not sure if my eyes caught my brain's memorandum. My eyes told a different story—see me, recognize me, am I yours?

Father O'Reilly wrapped his arms around me in a big bear hug and I knew the magnitude of the situation was too big to sweep under the rug. Father O'Reilly brushed back a tear from his eye while never dropping the intensity of my gaze and simply said, "Goodbye, Paul."

I fled to my car, fumbled for the keys and crumpled into the seat weeping like a small child. They were happy tears! They were tears of relief and tears of hopes and dreams colliding and mostly, tears of gratefulness. I had just spent the day getting acquainted with the man I knew was my birth father. Father O'Reilly was a good man. He was easy to confide in, carried a quiet confidence, exuded a strong sense of character and integrity, was strong in his convictions, showed a caring and sensitive nature, was a strong role model and overall, a very godly man. Yes, he had made some big mistakes, but I could sense something deeper. Then I remembered that the Bible says

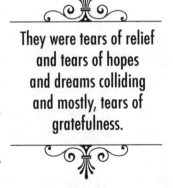

that I would be blessed if Father O'Reilly proved to be my birth father, *"The righteous man who walks in integrity and lives life in accord with his [godly] beliefs—How blessed [happy and spiritually secure] are his children after him [who have his example to follow]."* Pro. 20:7 (AMP) As I was dwelling on this I thought I realized that even though we were just really getting to know each other, I felt we shared many of the same qualities and characteristics, although I had a long way to go in the maturity department to match this man. If anyone was going to be my birth father, Father Ted was certainly the closest match of anyone I had ever encountered.

Through torrents of tears and exploding thoughts, I drove back to my hotel in Cambridge, pulled it all together long enough to check out and head towards Boston to return the rental car and jump on a train to Long Island, New York. After completing my search throughout New England, I decided to tie my trip into a visit to see my adopted parents and my brand new nephew Brendan, to whom my sister Karen had just given birth. The timing of this trip couldn't have been better; first my days with Father O'Reilly and now the new baby. It seemed God knew exactly what He was doing.

Chapter 9
THE PROBLEM WITH LIES

There are some things in this world you rely on, like a sure bet. And when they let you down, shifting from where you've carefully placed them, it shakes your faith, right where you stand.

Sarah Dessen, *Someone Like You*

The train chugged by at what felt like a snail's pace. It was a three-hour trip from Boston to Bridgeport Connecticut where I would then catch a ferry to Port Jefferson, NY on Long Island. Five long hours to replay the last week over and over in my mind. Images flashed before my eyes as my eyes fluttered with exhaustion.

I pictured my birth mother's house. I remembered the wild chase as I pursued the wrong car and then recalled Nancy's preposterous theory that night over dinner that Father O'Reilly could be my birth father. I thought about the hotel phone and my trepidation when using it to make big and scary phone calls.

Overall, I was thrilled to have discovered more substantive evidence in the search for my birth parents. I finally had proof of my birth mother's identity through the hospital medical records. And after significant quality time with Father O'Reilly my intuition was certainly telling me that he was my birth dad. But for all my productivity and investigative savvy, the emotional toll of the trip to New England was wearing me out, along with the lingering fatigue of my unexpected illness. I closed my

After I had come so far to meet her, I still couldn't believe she had reneged on our face-to-face encounter.

eyes again and considered the strange turn of events. First, there was the disappointment with Peggy. After I had come so far to meet her, I still couldn't believe she had reneged on our face-to-face encounter. But under the circumstances of secrecy and her shame of giving up her child, I understood on a more rational than emotional level.

Then the hospital visit, oh, boy! That poor woman still probably has her mouth open from the shock of me stealing her records. I laughed to myself under my breath, I still couldn't believe I did that. I could just see her telling her grandkids about the man who was so nice and polite and clean cut—who then turned out to be an imposter and a thief!

A bleak thought crossed my mind along with a flash of indignation: why hadn't my adoptive parents clued into this dysfunctional scenario? My fingers drummed on the side of the seat. After all, Father O'Reilly was their childhood friend and his sister is my godmother. My parents had grown up with Father O'Reilly for crying out loud! How could they be in the dark after all these years? Then an even more depressing thought tumbled into my consciousness, what if they were guilty of covering this up to protect Father O'Reilly and the young girl, Margaret Fitzpatrick?

I thought about the scarce details they had shared with me over the years and the decrepit orphanage in Brooklyn they pointed out to me on the way to visit family. They would say, "Paul that's where you came from." Later, as a teenager, I was told I was born in Cambridge, Massachusetts, when in truth I was born in Westfield, Massachusetts.

The more I thought about it the more irate I became. By the time the train pulled into the station, I was internally fuming! My hands clenched the reading table in my lap and my body tensed up thinking, I've spent the past three years searching and spending all this money trying to find answers when, perhaps, my adopted parents could have just sat me down and been honest with me.

Disappointment surged through me. I knew I needed to confront my mom and dad regarding this debacle immediately. It was a long overdue conversation. But then I thought about my sister's new baby and their elation at their first grandson. They were so excited. Did I really want to be the buzz kill on this celebration? But could I hide my angst? Could I conceal my confusion? I sighed with frustration. I guessed I would have to!

No one other than my wife Connie and my closest friends, Krystal and Rueben Joy, knew about my exhaustive search for my birth parents. I could call them if I needed backup. Their support bolstered me up and was crucial during setbacks. I thought how excited Krystal would be to hear all the details of this week because every time I saw her at work the first words out of her mouth were, "How's the search going?" Krystal had her own theories and ideas to solve this mystery and was almost as excited as I was.

I knew it would get my mom all worked up into a tizzy if I dropped this on her. And at this point, my emotions were so precarious I wasn't sure if I could handle hers as well as my own. So, I decided to wait on the confrontation. When I arrived in Port Jefferson, New York via the Long Island Ferry across the Long Island Sound, I took a cab straight to the hospital to see my sister and check out my adorable but very tiny nephew, Brendan. The next few days were a blur visiting with my sister, parents and childhood friends, Glen and Karen Petrosino.

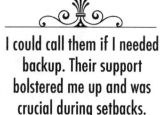

I could call them if I needed backup. Their support bolstered me up and was crucial during setbacks.

It was agonizing to stuff my emotions for the entire holiday weekend. Normally, I'm sad to leave my parents. But this time, I couldn't have been more grateful to board the plane heading home to California. I needed to digest and process all that I had learned. I also need to figure out my next move.

Chapter 10

A LACK OF EVIDENCE

Not knowing when the dawn will come, I open every door.

Emily Dickinson, *The Poems of Emily Dickinson*

In this mysterious saga of searching for truth, I learned to live in the uncertainty. Sure, I had theories about my biological father; in fact, I had strong suspicions based on highly logical assertions. But I still lacked the confession, the proof. If I were a prosecuting attorney, my lack of concrete evidence would lose the trial. My star witness, the man I believed was my birth father, refused to talk.

I was in limbo, making the best of a less-than-ideal situation; but forced to move on with the limited knowledge I had in hand. Dealing with real people and not the imaginary parents I had conjured in my mind was complicated. Reputations were at stake, lives built on fragile foundation. And the truth, so precious to me, was not as desirable to the other parties involved. The scenario was messy and I needed to tread lightly on my next step.

Once again, I drove to the beach, the one place I could retreat into my thoughts and reflect. I turned left onto Main Street in downtown Seal Beach and instantly felt comforted by the small town charm and the salty breeze blowing through my open window. I parked my car on a side street and headed down to the pier.

Fishermen lined the sides of the railing, but where the break crashed into foam and frothy white spray, an open bench beckoned. I sat down, and decided to write a letter to my godmother, Aunt Colleen. As I began to write, I realized for the first time, the woman I've called my Aunt Colleen for my entire life, because of the closeness of our families, may actually be my biological Aunt Colleen! I pulled out a notebook and began writing a letter to Aunt Colleen. Nine pages later, I looked up, shook my aching hand and read over the words poured out from my heart. The letter explained, in detail, the medical reasons for my search, including the recent diagnosis of my three-year-old daughter, Amanda with Juvenile Rheumatoid Arthritis. I clearly laid out the evidence I had assembled during my birth parents search, as well as my suspicions that Father O'Reilly was indeed my biological father. I knew this letter was a risk; it was a bold statement pointing the finger at a priest and her brother—it could blow up in my face. I was making a huge assumption that if Father O'Reilly was my birth dad, that Aunt Colleen was privy to the information. I was risking opening up a huge can of worms at an attempt to find the truth.

A few days later, the phone rang. It was Aunt Colleen and she had received my letter and was anxious to discuss the situation. Clearly, from her tone, she was in the dark and her brother had not confided in her. I could hear the shock and dismay in the cracks of her voice, but I also detected true curiosity as she considered the past, weighed the details, and put the pieces together. Aunt Colleen needed to hear more and she was open and supportive as I explained in depth my search for the truth. By the end of the conversation, I knew I had an ally in Aunt Colleen. She too believed her brother was more than likely my biological father. My theory made sense to her and connected the dots to questions she had struggled with over the years. I asked Aunt Colleen to keep this information quiet and refrain from telling her dearest friend, my mom, Flo Aubin. I wasn't ready emotionally just yet to share with my mom and dad about the search for my birth parents.

Aunt Colleen took my letter seriously and decided to contact her older sisters, Nancy and Mary, and share our suspicions. Apparently, they gave her enough insight regarding the strange circumstances swirling around Father O'Reilly during the time of my birth for Aunt Colleen to move forward and contact her brother. She set up a meeting to discuss the situation with him. Aunt Colleen strongly believed that if Father O'Reilly was indeed my biological father then he needed to come clean, because lives were at stake with the medical issues.

Aunt Colleen brought the letter with her when she met with her brother, Father O'Reilly. She emphasized the reasons for my search and my genuine motives, which were not to exploit or make a media frenzy, but to understand the past and build a relationship.

I don't know what Aunt Colleen said and I don't know the nuances of words spoken and unspoken, or even the dynamics of their relationship, but I hoped and prayed that something would tug at Father O'Reilly's spirit. I know in my own life, Aunt Colleen has always had a calming effect and her affectionate nature and soothing demeanor allow her to speak truth in a non-threatening and convicting way. Aunt Colleen is a woman of faith, and if anyone could tackle this sensitive and highly provocative issue, she could. She often quoted a Scripture from Ephesians when she was challenged to do good work for the Lord. *"For we are his workmanship, created in Christ Jesus for good works, which God prepared beforehand, that we should walk in them."* Eph. 2:10 (NKJV)

But while Father O'Reilly listened—he admitted nothing.

Their meeting passed and days went by, then weeks, and finally months. And so I waited. I waited and I got more discouraged—all this work and no tangible outcome. I had hit a large roadblock.

I waited and I got more discouraged—all this work and no tangible outcome. I had hit a large roadblock.

If my hypothesis was correct, Father O'Reilly simply had too much to lose—his reputation as a priest, his family's perception of him, and his credibility in the community. And there wasn't anything I could do to change his mind—it was in God's hands now.

Three months later—

I looked at my watch; it was almost go time. I was in San Diego for the launch of a new prescription drug for the pharmaceutical company where I worked. I gathered my notes, straightened my tie, and made sure I had everything I needed to rock my presentation. I looked at the time once more and figured I had a few extra minutes to go hunt down a phone to check my voicemails. My pager was vibrating on my hip and I had three new messages.

The first two messages were the normal business stuff. Then the third message began and I heard Father O'Reilly's voice. I pushed pause on the message, quickly shut the door to the pay phone cubicle, and exhaled, "Okay, Paul, here you go."

I started the message again. "Hi Paul, this is Father O'Reilly back in Massachusetts. It's been several months since we talked and I hope this message finds you and your family doing well. Paul, your Aunt Colleen shared the letter you wrote to her with me and I want you to know I was very impressed with the letter. It was very informative and well written and it must have taken a great deal of courage for you to write it and send it to your godmother." Father O'Reilly's voice paused.

"Paul, you deserve to know the truth about your history, given the medical circumstances and other reasons you highlighted in your letter. I understand your search has led you to believe I am your biological father. Leaving a message on your voicemail regarding this issue is less than ideal. I would have preferred to do this in person, but since I am clear across the country from you, this will have to do for now." Another long silence from Father O'Reilly, followed by a rush of words. "Paul, all of your suspicions about me are correct! I am your birth father!" Father O'Reilly's voice broke. "I'm sorry that it has taken thirty-five years for you to find out this important and sensitive information from your past. I'm sure it hasn't been easy for you, but I know you have been raised by two wonderful people who love you very much, my dear childhood friends, the Aubins."

I imagined he smiled through his tears as he thought of his good friends.

Father O'Reilly went on, "I know it will take some time for you to fully digest and process this information and it will probably generate more questions for you. When you're ready, I want to invite you back to New England to spend some quality time with you and walk you through the entire story from beginning to end. I have only one request: under no circumstances can you share with anyone outside of your immediate family because my wife and three children have not been informed of this. The sensitivities associated with this are endless and impact a lot of people. I hope you can respect this. Please call me at my private phone number so we can discuss this further and I hope to see you back in New England face-to-face real soon."

I slumped to the floor of the pay phone booth with a big goofy grin on my face and tears intermingled with relief. *I know, I know, I know* rumbled through my brain.

This was not another boring message I expected to receive on my voicemail. Elated with a blood tingling energy I jumped up and sprinted out of the booth to find my boss, Jerry. At the corridor I slowed my gait and walked up to Jerry. With an excited voice I said, "Can I speak with you privately?"

"Jerry, I found my birth father!" I eagerly whispered and then quickly explained the news, sparing him the details, but explaining why my head was not on straight with all the jubilance and shock of the moment. Fortunately, Jerry was understanding and scratched my presentation from the agenda. He encouraged me to take off the rest of the afternoon to let the news soak in.

I thanked him and hustled back to the pay phone to dial my wife, Connie and share the good news with her. Then I called my good friend Krystal, my Aunt Colleen, and then the private investigator, Nancy. They all affirmed my high spirits and celebrated with me as only people who are in the journey can. None of us could truly get our minds around the idea that a Catholic priest was really my birth dad.

I drove home up the coast from San Diego to Orange County in a daze. I was so shocked I struggled to contextualize my new reality. As soon as I slammed the front door, I hustled over to the phone and called for airline reservations to see what the costs would be to travel back to Boston and visit Father O'Reilly for a couple of days later in the month.

They all affirmed my high spirits and celebrated with me as only the people who are in the journey can.

In the morning, I picked up the phone and returned Father O'Reilly's call.

My voice felt almost childlike—so much excitement, fear, and trepidation intermingled with raw emotion spilled out of me. "Hi Father O'Reilly, its Paul."

"Good morning Paul," my birth father replied. "I want you to know how touched I was by your letter." Father O'Reilly paused and exhaled, letting go of burdens long held close to his heart. He then continued, "I felt obligated to share the truth with you about the past. It's time this secret was laid to rest. I also want to reassure you *son* that I don't have any cardiovascular disease and my cholesterol is well controlled with medication."

Father O'Reilly's voice lowered, "Paul, please remember how sensitive this information is. If the media catches a whiff of this scandal it will be splashed across every front page on the East Coast. The last thing I want is to see my family suffer because of my past. I don't want them to wake up and read the headlines, *My Father the Priest...* in the newspaper. Just to be careful, I have a private number I would like you to use, and when you leave me a message, the more generic the better."

I agreed to his stipulations, sensitive to the delicate nature of the situation, and then we reviewed our calendars and chose a couple of days later in the month to get together to go through everything life-defining! I was ecstatic to spend time getting to know my birth father and even more thrilled to finally put together the missing pieces of the puzzle.

Days passed—long anguishing days of waiting with jubilant expectation and finally, it was time. I boarded the plane to Boston, rented a car and spent the night in Cambridge before heading to Springfield the next morning. The plan was to meet up at Friendly's, our usual stop, and grab a bite to eat before heading to the Mohawk Trail again for a long drive with plenty of time to talk.

Father O'Reilly greeted me with his arms open wide—his embrace so different, but oddly the same. His smile was wide and his eyes crinkled at the corners with lines etched by both sadness and laughter.

The security of knowing he was my biological father changed everything. The moment felt weighty, elevating the simple ice cream shop to a place of significance. But although the magnitude of the reunion was great, our encounter wasn't awkward, despite my nerves and despite the anticipation. Father O'Reilly made me feel comfortable and the chemistry between

> Father Ted began to share the story as we drove down the Mohawk Trail—the story I had waited a lifetime to hear—the story of my birth.

us was natural and organic. His easygoing nature mirrored mine and put me at ease. It felt like home.

Father O'Reilly began to share the story as we drove down the Mohawk Trail—the story I had waited a lifetime to hear—*the story of my birth.*

Chapter 11

A Very Private Affair

If you're gonna tell your life story, you gotta be honest, or don't do it.

R. Kelly, *Soulacoaster: The Diary of Me*

Father O'Reilly started telling me my story, by first telling me his. "I met your birth mom at St. Andrews Church in Cambridge. I joined the priesthood in the mid-1950s and it was my first assignment. I was young and a little brash, in my late twenties, but Peggy was really young, just turning twenty. At first she worked as a volunteer at St. Andrews and then later moved into a full-time position in the rectory as an administrative assistant."

Father O'Reilly looked at me and grinned, "Your birth mom was lovely, Paul. She was tall and slender with golden blond hair and a knockout figure. That girl could light up a room, but it wasn't just her looks; she was smart and vibrant. She taught catechism and sang like an angel in the church choir. We worked closely together and were constantly thrown together in numerous church activities. Over time, our work friendship deepened and we became emotionally attracted—familiar and dear to one another."

Father O'Reilly gripped the wheel and leaned into the upcoming turn. "You see Paul, my family had traditions passed down for hundreds of years. There was a certain way things were done back then and I played the role I was expected to play. My parents emigrated from Ireland to New England after the turn of the century. It was a different world then, an almost archaic paradigm that seems strange in this modern world, but in my family, the oldest son was expected to go into the priesthood."

Father O'Reilly's eyes filled with tears. "Paul, the truth is, I fell in love with your birth mom. Our friendship turned into a secret relationship. We both knew we were playing with fire given the circumstances, but we justified our behavior because our feelings were so strong."

Father O'Reilly laughed sadly, "As if love, even true love, could overcome the fact that I was a priest. But, swept up in the fervor of infatuation, we couldn't see past our desire for each other. It was all above board, I mean nothing physical ever happened until 1958—the night of the youth retreat. That was the night," he turned and looked at me, eyes distant and reflective, "we—we crossed the line."

"There was a church youth outing, a charter boat trip from Boston Harbor to Cape Cod, and I was chosen as the priest for the trip and Peggy volunteered as a chaperone. Late that night, the phone in my cabin rang. 'Father O'Reilly, I'm scared,' Peggy said.

'What's going on?' I replied.

'I was in bed sound asleep and there was a loud knock on my door. It frightened me. I opened the door but no one was there. So, I climbed back into bed and started to fall asleep again and then someone banged on the door again. I jumped back up and ran to the door, threw it open and again, no one was there.' Peggy's voice reminded me of a child. 'I'll be right there,' I told her. I knew if I got caught in her room it would be the talk of the church. But the sound of her voice, her fear, her need for me, how could I not go?

"Once I got to her room and slipped in unnoticed, the knocking stopped. But then it was just the two of us, alone for the first time. And we were in a room with a warm and inviting bed and months of pent up desire. The next thing I knew she was in my arms and ummm, well—things happened."

Despite the fact we were both grown men with wives and children, I shrunk back in my seat, cheeks hot with embarrassment. I felt like a little boy and my dad was telling me about the birds and bees. It's always a shock to think about parents having sex—even weirder when you picture your birth dad in a robe and collar.

Father O'Reilly continued, "As the sun rose, my guilty feet scurried back to my private cabin like a mouse who had stolen the prized cheese. I was deeply ashamed of my behavior. How could I have broken my vows? The waves of guilt began to lap at my feet and with each step they grew in force and strength. By the time, I reached my room the full onslaught and ramification of my sin enveloped me in despair.

"Peggy, on the other hand, woke up with the exact opposite reaction. For her, the sun shined brighter, the birds chirped and she struggled to contain her euphoria. She was in love and reveling in the aftermath of our intimate encounter. Peggy hoped this was just the beginning of our romance. She didn't really consider the illicit nature of our affair. She believed I loved her and that was enough to make it all worthwhile. And I did love her, truly, but you see, I loved God more and I took my vows seriously, despite my moral failure.

"Paul, your birth mom was in love and her feelings were so powerful they eclipsed reality. I'm not sure she understood or cared about the consequences. She was young and as young people often are, blissfully unaware of the darker side of life. She didn't consider the people affected by our immoral actions—our families, the church, and our community would suffer because of us. They would feel the weight of dishonor."

Father O'Reilly's eyes looked off into the distance with sad resignation, "I *knew* better, but afterwards, I didn't know where to turn or whom I could trust with this information. I prayed to God and begged for forgiveness. Then I waited for a prompting from Him as to my next step. Should I confess to the church? Should I come clean and humiliate my family and congregation?" Father O'Reilly murmured in a low voice, almost a whisper.

"I decided to simply make it through the youth trip. I needed more time to process and digest the severity of my actions. I didn't know what to say or do around Peggy," Father O'Reilly sighed and shook his head, "I was torn and as she leaned in for reassurance, I emotionally shut her down. As long as kids surrounded us, I was able to keep it together. Peggy wore her best chaperone face trying to conceal a big smile, inadvertently making me feel even worse. We muddled through the return trip home to Boston Harbor and then on to St. Andrews Church in Cambridge. Finally, all the kids were returned back to their parents and I looked around and the church parking lot was empty. Now officially done and without the temporary distractions of the kids to avoid the situation, it was time to face the facts.

"So, I asked Peggy if I could give her a ride home so we could talk. I started the ignition and pulled out of the driveway onto the winding, woodland road. Peggy could sense I was on edge but she didn't understand why and she wanted to fix this distance she felt growing between us. Her hand rested lightly on my leg. I tensed up and she leaned in even closer, lips pressing into my collar to kiss and comfort me. I flinched and pulled the car over. I took her hand off my leg and turned to face her. Then I choked out a feeble apology, 'Peggy, please forgive me for letting things get out of hand. I got caught up in the heat of the moment, in how beautiful you are and your admiration for me was intoxicating.' I told her that I cared for her, but that I never should have taken advantage of her vulnerability.

"She sat so still while I talked, as if love itself was draining out of her. All the joy that lit her from within dimmed in those few minutes. A bright candle snuffed out.

"I finished talking and pulled back onto the road, stoic and determined to right this wrong—not realizing I was discounting her youth and emotions and heart.

"Paul, I expected more of myself as a priest. I was terrified of what this would mean to my parish, or for that matter, any other parish in my future. What would my parents say if they discovered I'd slept with a young woman who idolized me? They would have been mortified!"

I cried out to her, 'Peggy, you can't say a word of this to anyone, do you understand? Nobody!' I scared her Paul. She drew back as if my words were punches. I wounded her that day. Peggy's face fell and all her fairy tale dreams of us living together happily ever after came crashing down in the harsh truth of our illicit tryst. Her blue eyes turned to grey stone and tears burst forth in a torrent of anguish. 'But I love you and I thought you loved me. How can you do this to me?' she wailed as I turned into her parents' driveway.

"Before I could respond, she threw the door open and ran for her house. I called out to her but she never turned back. Then her mother opened the door with a confused look on her face and I got scared. I didn't think, Paul, I just reacted and pulled out of the driveway shaking like a leaf. Peggy's mom knew my car and it wasn't difficult to put two and two together and figure out I was the one responsible for upsetting her daughter.

"I couldn't sleep that night. I imagine Peggy didn't either. Her mom pressed and pressed for the truth but Peggy refused to divulge our secret, feigning exhaustion from the trip. She asked her mom to give her some space and much needed rest and her mother obliged, albeit reluctantly. We both knew if her mom got even a whiff of impropriety, she would march straight down to St. Andrews and demand retribution.

"The next day was brutal. I dragged my feet into the rectory knowing Peggy would be distressed and probably extremely angry. It was awkward to say the least. What I didn't expect was her reaction. She asked to meet with me after work and I agreed. We needed to talk more and put things in perspective. Peggy seemed distant and cold, jittery even when we sat down outside in a private wooded grove. 'Father O'Reilly, I think we should break off our relationship.' I could tell she had practiced those words all day. She paused and in that space I knew she wanted me to fight with her, to fight for us—but I didn't. The tears returned and in between sniffles she blurted out, 'You have made it clear there isn't enough room in your life for God and a woman. Father O'Reilly, I can't compete with God or the church. I wish things were different. This is so hard for me because... because... I love you so desperately,' and she broke down weeping. I tried to comfort her, but I have to confess I was so relieved. We could go back to being good friends and no one would have to know about our indiscretion. But I was kidding myself, Paul. It's not like these things go away.

"For a short time life returned to normal. Peggy and I kept our distance and hovered around each other, trying not to walk on eggshells, but the relationship had clearly changed. The easy-going friendship and flirtation disappeared. Clearly the sparks that drew us together and the magical days were gone although a big surprise lay just on the horizon.

"Another few weeks passed and Peggy started to feel physically ill. Suspecting she was pregnant, she made an appointment with a physician and waited anxiously for the results. After a few sleepless nights, a phone call from the doctor confirmed her fear. Peggy was pregnant with you, Paul—pregnant with the baby from a well-respected priest.

"I don't know how she worked up the courage to tell me. It must have been so hard! At some point after she found out she broke down and confided in her mother. I'm sure her mother was less than thrilled that her twenty two-year-old daughter was pregnant by me. In her heart, I know Peggy wanted me to leave the priesthood and marry her. She thought I could just get up and walk away from my vows and obligations. But it wasn't that easy, Paul. It wasn't a simple thing, "said Father O'Reilly.

"After procrastinating and stalling, Peggy's mom finally convinced her to take action and confront me. Peggy asked to meet with me privately one evening after we left the rectory. When she said, 'I'm pregnant,' I was floored.

"Here is this beautiful girl in front of me, weeping and pregnant with my child, no less, and I cared for her, I really cared for her, but it was so wrong," Father O'Reilly expressed with tears in his eyes. "I wasn't in a position to marry her. I was a priest. I needed to get right with God."

"I promised her that I would be with her every step of the way, but I reiterated how delicate a matter this was and the importance of concealing the matter for both her reputation and mine. If people found out about the baby she would be viewed as the Scarlett woman—the seducer of a beloved priest. I urged her to keep this a secret. We had to be careful," Father O'Reilly exclaimed.

"I prayed like I've never prayed before, Paul, with desperation and my pride shattered. Peggy, her mom and I decided it would be best to put you up for adoption. Abortion was never an option because of our beliefs. Peggy and I both knew the Scripture from Psalms that says, *'Your (God's) eyes saw my embryo, and on your scroll every day was written that was being formed for me, before any one of them had yet happened.'* Ps. 139:16 (CEV) You see Paul, once you were conceived we knew that God saw you as a living son of God and that He had already written down the days of your life even before you started living them. Having faith in the truth of this Scripture helped both Peggy and I during our struggle to face the fact that adoption really was our only choice.

Peggy was so young and being unmarried she couldn't financially care for you, and I clearly wasn't in a position to care for a child," Father O'Reilly lamented.

"Paul, I want you to know I cared very much about who would raise you. I approached my brother and sisters and confided in them, hoping and praying that maybe they would take you in and we could keep you in the family. But they all had large families of their own and weren't willing to take in another baby. They were concerned about trying to explain a random adoption to our traditional parents. I understood their reluctance, I mean it was complicated," Father O'Reilly confided.

"I chose not to share this with my youngest sister, Colleen, because she wasn't married at the time and I knew it would be upsetting to her. I also didn't want my parents to find out," professed Father O'Reilly.

"After each 'no' we were running out of options. We didn't want to put you in foster care within the church or worse, in the state-run orphanage of Massachusetts, but we were running out of time."

Chapter 12

THE COVER UP

The best way of keeping a secret is to pretend there isn't one.

Margaret Atwood, *The Blind Assassin*

Father O'Reilly continued to share his story as we drove further down the Mohawk Trail. "One day over dinner my youngest sister, Colleen casually mentioned a conversation with her good friend, Flo Aubin. I wasn't really paying attention. Truthfully my mind was racing with worry, until I heard Colleen mention the words *desperately wanting a baby*. I sat up and paid attention. Colleen explained that Flo had shared with her that she couldn't have a baby because of anemia. The Aubin's had been trying for years to get pregnant, but after multiple miscarriages the physicians had finally informed them that her blood couldn't nourish a baby through a full-term pregnancy.

"Colleen was sad for her friend, but I was secretly ecstatic. I knew that hearing Colleen's conversation was a message from

God. He was answering our fervent and desperate prayers. I excused myself as politely as possible, and ran for the telephone to call Peggy. I quickly called her and exclaimed, 'Peggy, the Aubins are desperate for a child.' Peggy was so relieved and elated that there was a couple she was familiar with who would raise their baby in a loving home. She also loved the idea that she could see you on a regular basis right here in Cambridge without anyone knowing. We agreed this would be a great choice for you, Paul. However, this wasn't an easy decision for either one of us. As the time came closer to give you up to the Aubins, Peggy became very emotionally distraught. She wasn't giving you up for a week while we were going on vacation; this was a lifelong decision that impacted all of us for the rest of our lives. We also decided it would be best if we didn't tell the Aubins where the baby came from, because if it got back to the church, it would be destructive for both of us.

"I know this probably sounds secretive and manipulative to you, Paul, and it was, but we were operating in a different world of societal norms. Unwed mothers were outcasts and considered vixens, the priesthood was sacred, and there was no room for moral failure. The system stood like a house of cards that was built on hiding and isolation. It was broken. There was no restoration or room for grace," Father Reilly stressed. "Sin was covered. Shame stuffed under the rug. In that day and time, in the late 50's, we really had no option but to find you a loving family. God placed

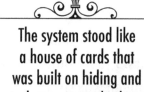

The system stood like a house of cards that was built on hiding and isolation. It was broken. There was no restoration or room for grace.

the Aubins in our path and His intent was so obvious regarding what we needed to do. My next step was to contact my sister Colleen and inform her that I would be willing to assist the Aubins in their desire to adopt a baby. Within the week, I spoke to Colleen and told her, 'I often visit a home for unwed mothers in Westfield. If the Aubins are interested

Sin was covered. Shame stuffed under the rug.

I might be able to help them adopt a baby. I might be able to do them a favor and put their name on the top of the list.' When Chris told this to the Aubins, they were thrilled! They believed this was God answering their prayers! Boy, were they right! God was answering all our prayers. They had no idea what a gift they were to us. The Aubins were like family and we knew we could see you when they visited Chris. We knew they were good people and you would be raised in a loving home nearby," Father O'Reilly explained.

"I asked my brother, Jack and my sisters, Mary and Nancy not to tell the Aubins or Colleen. I feared it would get back to my parents. I just couldn't risk it," Father O'Reilly added with a sigh. "Colleen was so close to the Aubins and I didn't want to put her in an uncomfortable position. She was very strong in her faith and so if she knew the truth, it would be extremely difficult for her not to share it. Mary, Nancy, and Jack had much less contact with the Aubins at that time."

As Father O'Reilly confessed, his countenance grew lighter as if, heavy bricks of sadness were being unloaded from his soul. I felt bad for him; this was clearly a very difficult conversation. I truly couldn't imagine carrying a burden like his for so long.

Father O'Reilly continued, "I contacted Frank and Flo in the summer of 1958, and told them there was a young woman at the home for unmarried mothers who was expected to give birth to a baby in October. I told them they were first on the list and to start planning for the big day. Paul, they were so excited! No baby was more wanted than you!" Father O'Reilly emphasized loudly with a pat on my back.

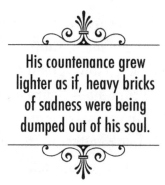

His countenance grew lighter as if, heavy bricks of sadness were being dumped out of his soul.

I had no words in response except a big "WOW."

Father O'Reilly continued, "By this time, your birth mom was starting to look suspiciously round and the risk of people gossiping grew daily," Father O'Reilly face grimaced. "It was becoming more and more difficult to hide her blossoming figure because of her slender frame. So, I made arrangements for her to move into the home for unwed mothers. The home for unwed mothers was about an hour away—close enough to visit, but, remote enough to remain undetected. I secretly visited your birth mom on my daily rounds and although I was gone more than normal, no one said anything to me about my absence. Peggy's mom also drove out each week to support her daughter through the last few months of her pregnancy. The secrecy was hard to maintain, but somehow we managed to keep the pregnancy concealed, even from from Peggy's dad—I don't know how, but he remained in the dark," uttered Father O'Reilly.

"But there were certain people at the church who were always skeptical of our relationship. We spent a great deal of time together, got along well, and despite our newly placed boundaries we still cared about each other. Your birth mom was a beautiful young woman, Paul. She was attractive and assertive and I was not so shabby looking back in the day," Father O'Reilly quipped. "We cut quite a figure together and people noticed."

"In the later months, I had to find a physician to take Peggy to for her regular checkups. We picked a doctor near the hospital where she would deliver. No one recognized me, so I signed all the paperwork as *Frank Aubin* knowing he would be your adopted dad," Father O'Reilly divulged. "The doctor confirmed the due date as October, but we didn't know what gender you were back in the pre-ultrasound days. We just had to wait and find out.

"On the morning of October 14, on her actual due date, strangely enough, Peggy went into labor. I drove out to pick her up, practically carried her down the stairs, and then took her to Noble General Hospital in Westfield, about five minutes away." Father Ted continued.

I choked back my laughter thinking about my crazy adventure with the old woman and the feeble security guards at Noble General Hospital, but that was a story for another day with my birth dad.

"After ten grueling hours of labor, Peggy gave birth to a healthy baby boy—that would be you, Paul," said Father O'Reilly with a little laugh. "You were born at 6:42 p.m. with no complications, other than a looming departure from your birth mother's arms. The original birth name we picked for you was Michael Patrick Fitzpatrick."

I nodded and checked off another piece of the puzzle. This was the name on the birth certificate I confirmed with my little heist at Noble General Hospital. Michael Patrick Fitzpatrick.

"Of course, Peggy built a bond with you in the hospital. She struggled to let you go. It was a natural bonding and she had a big heart. Paul, no mother should have to make this choice and I felt terrible—both conflicted and ashamed. She pleaded for just a few more days with you before she had to hand you over to the Aubins for good. But her mother was adamant. She told me, 'Contact the Aubins, immediately, before Peggy becomes any more emotionally attached to the baby.' Peggy begged for more time with you, but her mother and I assured her that you were going to a wonderful and loving home. We reminded her that she was doing the right thing and blessing this couple with her sacrifice. I would try to comfort her by reminding her of the Scripture from Psalms "*He (God) gives the childless woman a family, making her a happy mother. Praise the Lord!*" Ps. 113:9 (NLT)

Father O'Reilly's voice cracked, "It was a very difficult time for your birth mother. The reality of giving up her first child, the broken dreams and the cold hard truth—these were awful gut-wrenching and emotional days for Peggy. And she was never really the same. It broke her.

"I'll never forget that day. A chilly fall mist settled over the trees, as the dawn peeked through the clouds. I drove over to Peggy's house lost in my thoughts. I knew it would be a tough day. Although the drive was brief–the Aubins lived only a few minutes away in Cambridge–it was one of the longest trips of my life. The magnitude of what we were doing weighed heavy on my heart. I expected to see tears of joy pouring down the faces of the Aubins and tears of utter and complete heartbreak from Peggy. All I could do was pray for strength, Paul," my birth dad whispered.

"Peggy insisted that she ride in the car with me to drive you to the Aubins. I didn't think it was a great idea but she wouldn't take no for an answer. She wept the entire way there but managed to compose herself just before we pulled up to the Aubins' home. 'Peggy it's time,' I told her. Her eyes appeared dull and forlorn, like empty blue buckets drained of water and life. She kissed you gently and whispered her final goodbyes, then handed you over to me and I slowly carried you up the walkway to the front door while she remained in the car.

The emotional agony in the car transitioned into jubilation at your arrival. A little party had gathered—Frank and Flo, their brothers and sisters with spouses (your new aunts and uncles), your adopted grandparents, your godmother Colleen, as well as, additional family and friends. They were all gathered in the house awaiting my arrival with you. They were ecstatic, weeping tears of joy, especially your mom, Flo. You were the baby she had dreamed of and prayed for. Despite my guilt and shame and all of the brokenness of the affair, in that moment I felt redemption—God's hand at work. And a Bible verse came to my mind, *'And we know that all things work together for good to those who love God, to those who are the called according to His purpose.'* Rom. 8:28 (NKJV)

"From that day on your name was Paul Francis Aubin." Father O'Reilly declared.

"I basked in Flo's happiness, which felt very deceptive. After thanking me profusely, for bringing you into their lives, Flo asked me if I could baptize you."

At that point in the narrative, I was struck at the bizarrely awkward scenario. I thought, so, my own birth father baptized me. That's pretty awesome!

Father O'Reilly continued, "We met at St. Andrews Church later on in the afternoon for your baptism. It meant a great deal to me that I was a part of dedicating you to the Lord, Paul." Father O'Reilly revealed.

"But while the Aubin family was rejoicing over the arrival of their new son, Peggy was devastated. Her mom and I did our best to comfort her, but it was tough. Honestly, I was struggling too. I tried to stay strong, but it wasn't easy giving up my one and only son to another family. As a priest, I imagined I would never have children. All of that changed the instant you were born. My emotions raged against my logic. The next few weeks were hard days. I grew closer to God and spent hours on my knees, crying out for mercy.

"But time heals, Paul, time heals," Father O'Reilly exhaled. "And once everyone began to calm down, I knew we had made the right decision—although it was gut wrenching, to say the least. As the weeks passed, life resumed to a degree of normalcy, but nothing was really the same. People talk and it was damage control time. Tongues were wagging in the congregation. Whispers and pointed fingers hovered like an invisible cloud. This was front-page news and a scandal that would rock our town if the truth was ever to leak.

"Oh Paul, I lived in paranoia," Father O'Reilly recalled. "Would someone put two and two together and figure it out? Would they think—hmmm? Peggy took a leave of absence at St. Andrews Church for several months after gaining a few pounds around the middle and Father O'Reilly mysteriously disappeared early every afternoon? Fortunately, the Aubins were moving to Long Island,

New York for Fran's new engineering job, breaking their regular connection with anyone at St. Andrews.

"After several days of fasting and praying I knew I would be breaking my vows as a priest if I continued to conceal the truth and cover it up. I decided to speak with a mentor, a fellow priest at St. Andrews Church. He was a man with years of wisdom and I believed he could offer wisdom and guidance. After my mentor's initial shock, he strongly encouraged me to come clean and seek forgiveness. So I scheduled an appointment with the local bishop to share my predicament. Not surprisingly, the bishop was astonished and very disappointed to hear my confession of getting a young woman of the church pregnant. He pretty much torched me—not that I didn't deserve it. He let me know in no uncertain terms that as a priest and leader of St. Andrews Church I should have exercised more discipline with boundaries and never put myself in a compromising position. He told me I had one of two choices. The first choice was to leave the priesthood dishonorably and lose the respect of my congregation. The second choice was to serve in the Vietnam War as a chaplain on the front lines.

"My first thought was, what are my other choices? What about grace? But the grace didn't come. I was disillusioned and stunned by the two choices. Wasn't giving up my child enough? I was already hurting and now I was forced to make a life-changing decision. Leaving the priesthood dishonorably was unthinkable. Joining the army and going to the front lines in Vietnam to fight in the war sounded like a possible death sentence. The bishop gave me a few days to think about my decision, but I immediately knew what I had to do—and it didn't involve disgracing my family, Peggy, and the parish. I told the bishop I would enlist in the Army the very next morning. Over the next several days I notified Peggy, my family, and the people of St. Andrews of my intentions. And they, of course, were baffled. I didn't give them any specific reason for what seemed a very irrational and quick decision. I'm sure my decision was confusing to people that didn't understand the gravity of the situation. Looking back, I could have handled it differently, but I was reeling from my new reality.

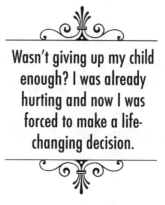

Wasn't giving up my child enough? I was already hurting and now I was forced to make a life-changing decision.

"Rumors and speculation swirled around the parish—tentative at first and then like a tidal wave of disgrace. Many of the women assumed Peggy must have had something to do with my sudden departure from St. Andrews. Peggy, unfortunately, became the scapegoat as the blame turned on her. As a result, Peggy was eventually let go from St. Andrews.

"Paul, I didn't know it at the time, but I left your birth mom in a terrible spot. She carried the entire burden of the debacle. The community viciously scorned her, which was exactly what I hoped to avoid. Peggy was shunned and vilified as a promiscuous vixen that ruined the life of their young, handsome, and beloved priest. In their eyes, I was the innocent victim, the sacrificial lamb, and she was the seductress. I didn't know the mess I was leaving her in.

"Peggy floundered in the wreckage of her life and her girlish dreams of romance came to an abrupt end. She was confronted with the emotional trauma of giving up her child and the ensuing social ostracism. While I was forced to deal with real bullets and attacking troops, hers was a far more insidious enemy. She was very young and didn't have the coping tools to deal with the overwhelming feelings of guilt and shame that engulfed her. It eventually led her into a deep depression that lasted for many years. The once poised, self-assured, and vivacious young woman that initially turned my head had transformed into a hollow shell. She eventually succumbed to the darkness and suffered an emotional breakdown requiring counseling for the remainder of her life.

"In the meantime, your adopted parents moved from Cambridge, Massachusetts, down the coast a few hours to Long Island, New York. Your dad accepted a job with an aerospace company as an electrical engineer. Your adopted parents returned to Cambridge about every other month to visit with their family and friends, and I always looked forward to seeing you and getting updates on your growth and progress. You sure were a tiny thing in the beginning and then, bam you grew so tall!

Father O'Reilly glanced over at me, took one hand off the wheel and patted me on the arm. "The Scripture that I've always prayed for you, Paul is from the book of Jeremiah, *'I know the thoughts that I think toward you, says the Lord, thoughts of peace and not of evil, to give you a future and a hope.'* Jer. 29:11 (NKJV) I believed that you would have a good future with the Aubins. It was a good fit for you, the Aubins, I know they love you very much."

I nodded my head yes, brushing away the tears fighting to break free. I thought of my mom, Flo with her big heart and endless generosity. I silently thanked God for this woman's unconditional love. And it hit me in that exact moment, as my birth father unfolded the layers of my story, adoption isn't the worst thing-separation is. I was blessed with the most priceless gift of a loving family relationship. But my birth mother, on the other hand, was destroyed by the other side of that same event. Her brokenness from a forced separation led to more hiding and isolation. She lived in a cage afraid to fly, afraid to love and terrified of ever being known.

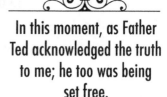

Adoption isn't the worst thing—separation is.

Although Father O'Reilly had his secrets too, he carried less of a burden, and it showed in his genuine warmth and spirit. I knew it was the mark of forgiveness. With Father O'Reilly's confession to God, his siblings, his mentor, and even the bishop, he had moved step-by-step towards restoration and healing. Yes, he still guarded the secret to avoid scandal, but today, in this moment, as Father O'Reilly acknowledged the truth to me, he too was being set free.

In this moment, as Father Ted acknowledged the truth to me; he too was being set free.

I said, "I remember seeing you on occasion when we visited my godmother, Aunt Colleen—your sister. And we spent time at Aunt Colleen's mother's home, just around the corner. I distinctly remember her thick Irish brogue and how kind she was to me." I sighed with a big whoosh of realization, "Oh right, she's your mother too—my grandmother." I chuckled.

I felt like Alice gazing at herself in the looking glass. What I knew to be true was a mirage. The players in my drama were still actors in the play; but the roles had changed. My godmother was actually my true aunt. I had called her Aunt Colleen all my life, but I didn't know she was truly my blood relative *aunt*. Someone

I thought was only a family friend was actually my biological grandmother, and my favorite priest was my birth dad.

An image flashed through my mind of the day at Babylon docks when everything changed. I remembered the harbor full of fishing boats, the sickening stench of clams, the sticky coffee ice cream at Friendly's, and the moment I found out I was adopted. And here I was full circle with all the questions finally answered. The missing pieces of the puzzle to my identity and family background were there. The wanting feeling, the ache I lived with my entire life—the void of not knowing my past—was finally gone.

The wanting feeling, the ache I lived with my entire life–the void of not knowing my past–was finally gone.

From the day she told me I was adopted, my mom had never stopped reminding me of how special I was to her, how much she and my dad loved me and how nothing else really mattered. But, I don't know if I ever fully believed it. The doubts outweighed the certainty. But today, thirty-five years later, I finally knew her words were true. I also knew that although many of the players had shifted roles, my mom had not. The woman who raised me was the true heroine in this drama.

I shook my head trying to make sense of it all. I looked at my birth dad driving, and smiled at his big hands just like mine. I sat back in my seat, closed my eyes and relaxed in this new space of knowing.

Chapter 13

HIDDEN IN PLAIN SIGHT

I thought about how there are two types of secrets: the kind you want to keep in, and the kind you don't dare to let out.

Ally Carter, *Don't Judge a Girl by Her Cover*

A few months passed as I got used to my new reality. I stayed in contact with my birth father and our tentative relationship blossomed. But the more I appreciated the love of my heretofore unknown birth father, the more frustrated I got at my adopted parents lack of transparency. The grace I initially bestowed upon them was starting to turn into reasonable questions. How could they not know the details of my adoption? Were they simply naive or did they know far more than they led me to believe?

Surely they guessed, or at the very least questioned, if Father O'Reilly was my birth father since they were friends with him dating back to grade school. Couldn't they see a resemblance?

How could Father O'Reilly conceal from his dear friends that the baby they were adopting wasn't a stranger's baby?

My logical and analytical side chewed on every pinhole in the details. It just didn't add up. My adopted father's name was on the original birth certificate that Father O'Reilly and Peggy filled out. Didn't my parents ever question the absurdity? How would an unwed mother know where her baby was heading after releasing them from the home for unwed mothers?

From my earliest memories as a boy, my adopted parents had led me to believe I was born in Brooklyn and that it was there, in the big, scary building full of sad orphans that they had picked me up. Clearly, this was a falsehood. I could understood why they might refrain from sharing these details with me as a small child, but why continue to hide the truth from me well into my thirties? My frustration grew the more I dwelled on it and I became indignant at the cover up concealing the truth of my past and the flat out lies over many of the details.

I decided it was time to confront them and let them know the results of my search. I flew back to New York with a heavy heart. They met me at the airport and I struggled to remain composed—honestly, I struggled to simply be pleasant. After a visit with my sister and her family in Port Jefferson, we drove back to my parents' house in West Babylon. As we pulled into the driveway, sweat trickled down my brow. It was time to face up to the truth.

I asked my parents to join me in the dining room while explaining I had something sensitive to share with them. My mom looked nervous and started to fidget, folding and refolding the napkins and my dad seemed anxious and tense. I knew they were concerned about my health. I fought back my rising irritation. On their behalf—they probably thought this was a bad news conversation. No one likes to hear, we have to talk. They were already aware of my diagnosis with high cholesterol three years earlier, and I had divulged that my cardiologist strongly recommended I search for my birth parents. They knew I needed to gain important insight into my birth parents' medical profile so my cardiologist would know how to treat my condition. If my birth dad suffered a heart attack at an early age then he would treat me more aggressively, but if he had high cholesterol managed by medication then Dr. O'Connor would feel more comfortable administering a cholesterol lowering drug and monitoring my diet.

As soon as they sat down, I dove in "Mom, dad—I found my birth parents!"

Whatever I expected, it was not this! Silence. Crickets. Big eyes staring back at me.

I plunged on despite the awkwardness and informed them, while their mouths were still gawking, of my three-year journey to find my birth parents. I waited for their confession, but none came.

The shock on their faces appeared sincere. Although I had divulged very little about my search to them, it seemed that maybe they were in the dark too.

Now slightly confused at their response, I rambled on. I shared with them how I employed a private investigator specializing in finding birth parents for adoptees, and after three very long years I had, in fact, found my birth parents.

My mom's eyes filled with tears and my dad appeared dumbfounded. They were absolutely stunned that I had navigated the search for my birth parents over the past several years without informing or including them. I could see the hurt in their eyes and a sliver of guilt shot up my spine. I tried to explain my reasoning—that it was enough of a volatile emotional roller coaster for me to experience without putting them through it as well. I reminded my mom of her tendency towards anxiety and how I was trying to protect her.

My mom nodded her head in agreement, tears welling up and sniffles escaping, "I guess I do get a bit nervous. But, Paul, I feel terrible you had to go through all of this on your own."

"So, son, what did you find?" My dad Frank leaned in. I could sense their unease and heightened apprehension. They were chomping at the bit, eager to find out who my birth parents were. But I wasn't ready to share the big news with them until I had some answers of my own. Like a little kid with a toy I didn't want to share, I waited and let the tension build.

I took a deep breath, "Why weren't you honest with me from the beginning or at least when I got into my thirties?" Taut with hurt, my voice sounded stiff. "What I uncovered wasn't even close to what I was led to believe all these years."

My dad tensed up, fume erupting from his nostrils, his normal reaction when he felt out of control. My mom began to cry, "Please, Paul, believe me, I told you the little I knew throughout the years but we had very limited information."

"Okay," I began, "Here's how it unfolded..." And I painstakingly walked them through the details of my search starting from the cardiologist and initial diagnosis to finding Nancy and onward. My dad's breathing became hard and labored and my mom shook—tremors and nervous sighs erupting in between fits of tears.

Finally, I dropped the bomb, "Mom, Dad…Father O'Reilly is my birth father."

"What?" they choked out together. Their eyes met. Hurt poured out my mom's inky wet lashes and hatred emanated from my dad.

"I'm sorry, it's true," I faltered and then spoke more softly. "Your long time friend, Father O'Reilly, is my birth father and a woman who worked with him at the church named Peggy Fitzpatrick is my birth mother."

Their reaction assured me they didn't know Father O'Reilly was my birth father. My mom buried her face in her hands and wailed while my dad looked furious.

My dad, Frank viciously lashed out, "He lied to us to protect his own skin! How could he deceive us for so long? Father O'Reilly, the war hero and the priest fathered my son? Give me the keys. I'm going to drive up there right now and confront him for lying to us—for lying to the world."

"Please Frank, think about Paul right now, and stop your tirade!" My mom reached down and grabbed my hand. "I'm so sorry Paul," she murmured and then opened her arms to hold me.

I leaned in and smelled her familiar perfume. Like a small boy I held her close, nestled and safe, as tears coursed down my cheeks from weeks of pent-up fury. I looked in her eyes and behind her weak smile, I saw fierce and raw pain. She was wounded and I knew, without a doubt, they too were victims in this debacle.

"This must have been quite an ordeal for you to go through Paul," Mom paused and a look of dismay crossed her face. "I am so disappointed in Father O'Reilly. We have worshipped that man and held him up on the highest pedestal for giving us the child we weren't able to have on our own. His was the ultimate gift and blessing that I could never repay. Now to find out thirty-five years later that his gift wasn't genuine but self-serving! He had a hidden agenda all along; to protect himself!"

Over the next few days their disappointment turned into a festering bitterness—the betrayal of Father O'Reilly ran deep—and the debt they felt they could never repay turned in on itself. Now Father O'Reilly was the debtor and they held the cards—tightly.

The remainder of my trip was painful. Each day was a new opportunity for my adopted parents to probe me for more details from first contact down to the last conversation. My parents, not surprisingly, felt cheated and resentful. In some ways I wish I had kept it to myself because once out in the open, there was no going back. I thought I would feel better with everything out in the open, but strangely enough now there were new issues. I felt caught in the middle of their animosity because I liked Father O'Reilly and I wanted a relationship with my biological father. I had extended forgiveness they weren't ready to dole out. It was clear, at least right now; while emotions were hot I wasn't going to find any encouragement from them regarding my biological dad.

Chapter 14

KEYS FROM THE PAST

And above all, watch with glittering eyes the whole world around you because the greatest secrets are always hidden in the most unlikely places.

Roald Dahl, *The Minpins*

Near the end of my trip, a few weeks later, I headed back to Cambridge for a visit with Grandma Aubin. She was my adopted dad's mother. Grandma's health was ailing and she looked a little frail to me, but her spunky nature was still very apparent. I wrapped my arms around her and settled into a chair next to her in the living room. We caught up with each other about all the topics grandma's and their grandsons normally talk about. Then I told her about my search for my birth parents. I held her hand, trying to memorize the details of her laugh and quick smile because I knew this might be one of the last times I would see her.

Just as I was getting ready to depart, I leaned in for a goodbye kiss and Grandma Aubin murmured, "You know Paul, I knew what happened from the very beginning."

"Grandma, what are you talking about?" I queried.

"Well, I was sitting on a bus one day, many years ago, when I overheard two ladies in front of me gossiping about the young priest at St. Andrews Church who suddenly left the church without notice, the rumor going around town was because he was having a secret affair with one of the young women who worked in the rectory." Grandma gasped a bit, took a deep breath and trudged on, "They said she got pregnant by the handsome priest and then had a baby boy whom she gave up for adoption to a local couple that moved to New York.

"Paul, I'm not stupid. I sat in my seat quiet as a mouse and thought about it. It didn't take me long to put all the pieces together to determine that the baby boy they were talking about was, in fact, you my grandson. But I never said anything to your parents. It didn't seem like my business to interfere." Grandma gave me a weak grin and closed her eyes needing rest. "Now, you go and be safe. Grandma loves you."

I walked out of the room, gently closed the door and laughed to myself in disbelief. I couldn't believe what I had just heard. I could have saved a huge sum of money and years of frustration if I had only visited my ninety-three-year-old grandmother and solicited her wisdom and perspective! But, the more I thought about it, the more grateful I was for our encounter. Seeing the incident from her point of view helped me to understand the vicious gossip my birth mother endured. It must have been really vicious, if people were boldly tittle-tattling on the bus. The rumormongers must have been in full swing back in the day. I was grateful my parents moved to New York, which stilled the gossipers and I was glad I didn't have to grow up under a cloud of disgrace.

This outing to see my grandmother was unfortunately my last visit with her before she passed. I was deeply grateful for the opportunity to say goodbye. What a gift my grandma left me—simple clarity.

I boarded the plane back to California and reflected on my trip. Confronting my adopted parents was messy and I'm not even sure I discovered the whole truth, but at least I'd dug up their version of the past. Somewhere, in the midst of the real-life drama between Father O'Reilly and Peggy, Flo and Frank was my reality—although trying to ascertain exact accuracy proved elusive. The details of the past always seem to be swayed by point of view.

The details of the past always seem to be swayed by point of view.

Upon my arrival home, I scheduled an appointment with my cardiologist in order to share the findings of my journey. I climbed up onto the examination table, put my feet up on the crinkly paper, and asked him to take a seat because it was a whopper of a tale.

Quickly, I summarized the details of my search and Dr. O'Conner shook his head in disbelief. He laughed at my Noble General Hospital exploits, widened his eyes when I revealed my birth dad was a priest, and then leaned in when I described my birth father's health.

"You were right Dr. O Conner," I said. "My cholesterol appears to be a genetic condition. My birth father has extremely high cholesterol also, but he hasn't suffered from any heart disease and he's in his late sixties. That's good right?"

The doctor nodded yes. "What medication is he on?"

"It's called Mevacor," I replied. "And it seems like it's doing a decent job of keeping his cholesterol levels within normal limits."

"Paul, I'm stunned by the journey you have been on over the last few years. I don't know anyone who's had to jump through so many hoops to get a health history, but at the same time, I'm very pleased at what you've discovered. I'm going to prescribe the same medication that your birth dad is on, and recommend you continue with diet and exercise. And Paul, let's try and keep the stress down too," he winked with a playful grin.

Chapter 15

A Beautiful Shell

Lies and secrets, Tessa, they are like a cancer in the soul. They eat away what is good and leave only destruction behind.

Cassandra Clare, *Clockwork Prince*

In January of 1996 my wife, Connie, and my daughter, Amanda and I packed up our home and moved across the country from Southern California to Cary, North Carolina. This move was due to a new position that I was offered within the pharmaceutical industry at Glaxo Wellcome, Inc. in Research Triangle Park, North Carolina. I was sad to say goodbye to the sunshine, family, and friends, but excited for the new work opportunities and the progression of my career. Another unexpected perk to the move was our location. By moving back to the East Coast we were able to easily connect with my family and old friends in New England and New York. And, as I established a renewed relationship with my birth dad, it didn't hurt to be a little closer.

I found myself back in Boston one evening, wandering around the historic downtown area looking for a first-class Italian restaurant. I was craving pasta and Boston has some of the best places around. I was exhausted after another long day in a stuffy conference room. I was the Director of Strategic Alliances at Glaxo Wellcome and they were sending me off to Harvard Business School for some additional education.

After a few blocks of not finding what my nose was looking for, I stumbled across a trendy Italian café with an inviting outdoor patio. Tempting smells drifted through the door and I spotted a lone table with a stunning view. I requested it and settled in. I dove into the menu with delight and ordered an impressive looking half bottle of Merlot. The food was exceptional, maybe one of the best Italian dishes of eggplant parmesan I've ever experienced, and I thoroughly reveled in the ambiance. As I sat on the patio, overlooking the city lights of Boston, I began to reflect on my adoption search, running through details and processing events.

The what ifs played a tune in my head, life would be different, but would it be better?

What if I hadn't been adopted? My life would certainly look entirely different. There was even a good chance I might be working here in Boston and living in the city or in the suburbs. The *"what ifs"* played a tune in my head. Life would be different, but would it be better?

Images of my birth mother ran through my head. I wanted to meet her. I craved the knowing of who she was and what she was about. All I had were fragments of stories from people reminiscing about a young and scared girl. Who was this woman?

It could have been the wine or maybe the delicious food, but it suddenly struck me with an overwhelming veracity that my birth mother was only an hour away in Worcester, and I was no longer clear across the country. My hands were no longer tied by distance. Why shouldn't I take advantage of my time here by reaching out to her one more time?

The doubts and fears of her possible rejection again fell away in the surge of adrenaline that coursed through me. A few years had passed since I had suddenly revealed my identity. Maybe now she would be more approachable. Maybe the seeds of curiosity about her son had rooted and now maybe she would be more open to meeting me.

I paid the bill quickly and hustled back to my hotel room at the Marriott Copley Plaza to prepare for the phone call to my birth mother, Peggy. Over an hour later, after numerous and exhaustive rehearsals, I finally picked up the phone and dialed her number. I swallowed down the nerves and anxiety tumbling around with the pasta in my stomach.

"Hello," a young girl answered the phone.

Is this Peggy's daughter? I questioned myself. My half sister who doesn't even know I exist?

"Is Peggy home?" I casually inquired.

"Who is calling?" she replied sweetly.

"I'm an old friend of Peggy's," I quickly responded.

Painful seconds elapsed as I waited on the line trying to calm myself. Eventually, I heard voices and heels clicking on wood floors as Peggy came to the phone.

"Hello, Peggy said politely, "Whom am I speaking to?"

I said, "Hi Peggy, this is Paul Aubin."

There was long pause—"Oh hi," I could hear her hesitation in front of her daughter.

I felt like a telemarketer because I knew I had about ten seconds to get her buy-in before she hung up on me.

"I know when we first talked a couple of years ago, you were resistant to meet with me because the search for my birth parents was extremely sensitive and could potentially open old wounds and hurt people. I was, as you can imagine, disappointed you didn't want to meet with me, but I tried to respect your position and I haven't attempted to contact you for over two years. The purpose of my call today is to let you know that I no longer live in California. I have moved to the East Coast to North Carolina; and I'm actually right down the road staying in Boston for the remainder of the week on business. I was hoping we could take advantage of my time here and finally meet in person, over a cup of coffee, at your convenience."

"Paul, as much as I would like to, I'm really not comfortable with this situation," Peggy retorted.

"Well, Peggy, there is something you need to know. When you were unwilling to meet with me, I was forced to search for my birth father to gain the necessary medical information I required. And fortunately, I found him. Father Ted shared everything with me from the very beginning and he confirmed that you are indeed my birth mother. So—I would like to give you the opportunity to meet your son for the first time since you gave birth thirty-eight years ago in October of 1958." I took a deep breath and moved in for my final line of reasoning. I was hoping to force her hand. "If you still don't want to meet with me I'll never attempt to contact you again. Do you want to meet over breakfast or lunch tomorrow?"

"Well, I suppose I could meet you around 11:00 a.m. tomorrow in the lobby of the Cambridge Marriott," Peggy relented.

"Fantastic! I'll see you at 11:00 a.m. tomorrow morning." I confirmed.

I tried to contain my excitement, not wanting to face unnecessary disappointment if Peggy had a last minute change of heart, but it was difficult to sleep. My mind raced with rapid-fire images—impressions of a young, beautiful girl defeated by tragic circumstance, then a woman my age surrounded by kids and homemaking. Finally, I projected forward to the woman of today. I pictured her as a simple but lovely blue-collar housewife, minimal make-up or jewelry and dressed as a traditional, conservative New Englander. I imagined her to be similar to my adopted mom, vibrant, nurturing, talkative, and maybe a bit skittish due to the covert nature of our meeting.

What would we talk about? Would she offer more clues to the remaining pieces of my unsolved puzzle? Would she shed tears? Would I? I imagined her side of the story was poles apart from Father Ted's. I expected her feelings were still sore and distressing even after all these years.

I tossed and turned, pummeling my pillow. What will she think of me? Will I remind her of Father Ted? Will my resemblance to him strain our connection? My nerves warred with the great meal lingering in my belly and my stomach churned in anticipation. Meeting my birth mom for the first time at the age of thirty-eight was certainly dismantling my emotional equilibrium.

I woke up from a deep sleep, groggy eyed but energized. After a warm shower, a sizeable cup of coffee, and a light breakfast, I headed off to the car rental company. I felt guilty as I walked by Harvard's gates and kept on going but as I wasn't going to miss this once in a life-time meeting with my birth mother to attend a post-graduate class on strategic alliances. But I kept my head down, pulled up my scarf, and studied the sidewalk, just in case I ran into a curious colleague.

It was an archetypal New England day, rainy and cold with lingering ice on the road. I headed out of Boston and drove an hour east through curving roads and lush tree lined canopies to the Worcester Marriott.

I pulled into the parking garage of the hotel about twenty minutes early for our 11:00 a.m. rendezvous. I was eager to catch a glimpse of my birth mother before we officially met. I loitered in the car craning my neck for a peek as each car parked. I desperately hoped to see Peggy walk from her car to the lobby of the hotel.

What felt like an eternity passed, although it was probably only ten minutes, when a car pulled into the garage with what appeared to be a solitary woman? I peered into the rear-view mirror and watched the woman exit her car and sedately walk to the front of the Marriott hotel. As she strolled past the rear of my car, I caught the impact of her presence—late 50s, tall, slim, and blonde, in professional and elegant business attire. I sucked in my breath in befuddlement. This was a very attractive woman and not how I envisioned my birth mother to look. My first impression was, "no way could this lady be my birth mom! "I'll continue sitting here in my car until she eventually shows up." The

A Beautiful Shell

minutes ticked by and nobody else pulled into the parking garage. Now at 10:57 a.m. and close to our 11:00 a.m. meeting time, I decided to venture into the hotel and wait for her there. Perhaps there was another entrance on the side or the back of the hotel I had missed?

Rain battered the ground as I opened the car door and stepped around muddy puddles. I hurried into the hotel and an uneasy feeling came over me. Was it possible the tall, attractive, blonde woman who passed by earlier could be my birth mother?

As I crossed over the threshold and through the doors to the hotel, I quickly scanned the lobby noticing a few men and only one female—the woman who passed me in the parking garage. She was delicately perched alone on a chaise in the lobby. She was fiddling with her watch and her nervous demeanor reflected my own.

It can't be her! My brain battled with the obvious. I looked around for the restroom and practically ran through the lobby for a moment of respite. I turned on the water and splashed my hands and face, gathering my thoughts and willing myself to calm down. A wave of nausea tugged at my throat and dizziness forced me to grip the washbasin. "Pull it together Paul." I whispered to the reflection in the mirror. I slowly inhaled and then exhaled deeply, swallowing the overwhelming sensation to flee. Okay, time to face the music. I braced myself and then headed for the lobby knowing this

I braced myself and then headed for the lobby knowing this moment would be etched in my mind forever—a monument of inauguration—the day I met my birth mother.

moment would be etched in my mind forever—a monument of inauguration—the day I met my birth mother.

The chic woman from the parking garage was still sitting alone in the lobby looking as if she was anticipating the arrival of someone. At a snail's pace, I gradually walked towards her, feet shuffling with hesitation, as her inquisitive eyes met mine.

A wide smile crinkled her startlingly blue eyes, "Paul, is that you?"

I timidly grinned back and said "Peggy?" *This was it.* This was my birth mother. I stood frozen as emotion choked my throat. And then this beautiful woman, who was related to me, opened her arms and leaned in for a long hug—a lifetime of longing embrace—a hug for the boy and now man she never knew.

I folded like a child, hot wet tears streaming down my cheeks as Peggy clung to me weeping, black rivers of mascara streaking her lovely face. Her smell was bizarrely familiar—an instinctual connection I didn't comprehend but innately understood. The moment stilled, time paused, and then I felt the weight of eyes upon us and sniffs and throats being cleared beyond our own. Through my tears, I glanced up and noticed a small crowd of people watching. It seemed our reunion was a little less private than anticipated. The hotel clerks wiped away tears and then turned away with weak and shy smiles to give us privacy.

And then this beautiful woman, who was related to me, opened her arms and leaned in for a long hug—a lifetime of longing embrace—a hug for the boy and now man she never knew.

A Beautiful Shell

Apparently, while Peggy was waiting for my arrival she disclosed to the staff that she was meeting her son, for the first time in thirty-eight years. The excitement and commotion reminded me of the commercials of soldiers returning home and the shock and surprise of their families. It was a rare moment made all the more precious because of the intensity of emotion—a culmination of a story played out for a hotel audience to experience.

Peggy dug through her purse for a handkerchief, wiped her eyes and motioned for me to join her in the restaurant. As I followed behind her, I tried my best to pull it all back together. We picked an out of the way table and settled in, eager to get acquainted with each other. For the next three hours I stared into her blue eyes trying to catch up on the previous thirty-eight years—

Chills of wonderment slid down my spine; it was so peculiar to meet a relative, someone actually connected by genetics to me with such similar features.

no small feat! At first, our overwhelming sentiment precluded us from speaking; we simply sat motionless, mesmerized with one another, pointing out the resemblances. Peggy shared my Swedish features of a long, narrow face, bright cerulean blue eyes, elongated fingers, lean, lithe build, and matching smile. Chills of wonderment slid down my spine; it was so peculiar to meet a *relative*, someone actually connected by genetics to me with such similar features. I was enthralled by the experience—spellbound even—and couldn't stop staring at her from head to toe.

Once we exhausted the small talk, exchanging pleasantries and gazing goofily at one another, we dove into the more precise details regarding families, residence, careers, etc. Even though

Peggy freely shed light on her own life and peppered me with basic questions, I got the impression she was merely biding her time for a bigger conversation. My initial ease was dissipating as I picked up on an unspoken agenda to our meeting. I watched her face closely and looked for clues in her body language. I noticed tenseness around the corners of her mouth when Father O'Reilly's name was mentioned. She knew I had previously heard Father O'Reilly's side of the story. As the subject of Father O'Reilly was broached, I clung to the warm, fuzzy feeling of reconnection, as a frost of vindication appeared to settle over her demeanor.

As the subject of Father Ted was broached, I clung to the warm, fuzzy feeling of reconnection, as a frost of vindication appeared to settle over her demeanor.

"Paul," Peggy queried with a thin smile, "why don't you share everything Father O'Reilly told you?"

"Okay—ummm—here's what he explained to me," I stammered. Out of respect for Father O'Reilly and a growing sense of apprehension, I measured my words carefully and simply illuminated the highlights of my discussion with Father O'Reilly. I was trying to remain as neutral as possible.

Peggy nodded her head, "Yes, that's the gist of it. You know Paul, I never intended for this to happen and it wasn't some *little* thing that I was ever able to get over. It destroyed me for life! I've struggled every single day since I gave you up with despair and depression. In fact, the adoption haunted me. I didn't want to give up my baby. They forced me to! And I never intended to fall in love with a priest or for any of this to happen. I was just a girl who fell in love, a girl who no one was willing to fight for. Paul, they were vicious! The church—all those self-righteous people—they treated me like an outcast. I was shunned! I took all the blame and Father O'Reilly went off to become a war hero in Vietnam and chaplain at Arlington National Cemetery. He was the champion and I was the harlot! He took advantage of a young girl with a crush and annihilated me.

"The injustice enrages me Paul. How dare he pretend to be so godly when he used me and brushed me aside? I was ashamed to walk out of the front door. People whispered under their breath that I was the church whore—but Father O'Reilly, on the other hand, he came home to a parade."

Peggy's lovely smile turned into a sneer as bitterness simmered under the surface. "I've been in counseling for years to try and put this into perspective—but the guilt and the shame and the anger still consumes me at times. I had to get away and start over. I moved away from the gossip and I buried the past. No one knows about the girl I was; including my husband and children."

I tried to process her words and not react to the venom pouring out of her. How could she keep this scathing secret for so many years? How could she live in fear for so long? Years spent building resentment, feeling cheated, and resigning herself to concealing her own identity. Masked behind Peggy's beauty and her perfect makeup was an angry volcano ready to spew.

Wanting to lighten the mood, I reminded Peggy that I was also searching for information regarding my medical history. "Does anyone in your family have heart disease or highly elevated cholesterol?"

"No, we are a pretty healthy lot with no significant cardiovascular issues. Why are you so concerned about my cholesterol history?" Peggy asked with her brow arching.

I described in detail the critical discovery I made that day at the pharmaceutical convention and how it led to the cardiologist, more testing and then the pursuit for my birth parents to obtain the necessary medical information.

Peggy chuckled at my exploits, but frowned and tensed up any time I brought up Father O'Reilly. Her curiosity about him was obvious. She probed for details of his version of the story, but I felt uncomfortable breaking his sacred trust, as she clearly wanted me to. So out of respect for him, I dodged her questions and was intentionally vague in my responses.

Peggy then inquired about my daughter, Amanda—her biological granddaughter.

"Do you have a picture of Amanda? Does she look like you?" Peggy's smile gradually returned as I handed it to her. "Tell me about her."

"Amanda's amazing," I replied, happy to brag about my beloved daughter. "She's gorgeous, clever, and athletically gifted. She takes after her daddy and loves sports; she's an all-star softball player and a gifted golfer with natural ability. But best of all, Amanda is simply a good and kind-hearted little girl who never gives us a lick of trouble."

> I kept waiting for her to ask me a few questions about my childhood or upbringing, but to my disappointment the conversation continued to focus on Father Ted.

We talked on and on but Peggy kept returning the conversation to Father O'Reilly, launching into one tirade after the next on his inappropriate behavior. I kept waiting for her to ask me a few questions about my childhood or upbringing, but to my disappointment the conversation continued to focus on Father O'Reilly. In the several hours we spent together that morning reuniting, it became evident my birth mother was emotionally distant and not a very warm person. She seemed intently focused on communicating that she was the victim. She laid sole blame on Father O'Reilly shoulders, while never considering what I experienced all these years as a result of their joint decision.

It was obvious I wasn't a welcome reminder of the past. The affair with Father O'Reilly, my subsequent birth, the adoption, and the following chaos that ensued had clearly scarred Peggy for life. It had transformed her from a bright and vibrant young woman into an embittered and beautiful, but aloof shell. I truly wanted to give Peggy grace. I wanted to deeply connect and bond with her without it being forced. I wanted to feel like I did with Father O'Reilly, but something held me back—probably her veiled anger directed at not only my birth father but at the world. I didn't feel she wanted to know me the way I longed to know her—as a mother relates to a son. Mostly, I sensed she wanted an ear to vent to and a place to justify almost forty years of pent up rage. But, then again, I suppose having me show up in her life unexpectedly woke up the sleeping dragon she had concealed for far too long.

The hours flew by as we talked and all too soon our time slipped away. Peggy needed to get back to work and I couldn't put off my seminar at Harvard Business School any longer. We headed for the parking lot where we said an emotional goodbye, hugged again and agreed to keep in touch.

"Paul," Peggy whispered as she released me from her embrace, "please only contact me at my office because my family, as you recall, is not in the loop on this matter."

Her eyes looked panicky, "I promise I'll be sensitive," I replied with a nod.

Peggy gave me a faint smile and closed her car door, waving at me as she pulled out. It's the first and last time I would see her.

The drive back to Boston and Harvard was a blur. I didn't know whether to laugh or cry. My head ached and my hands shook on the wheel as I processed the encounter. The car bumped as I hit some gravel—"Oh man, I probably shouldn't be driving right now," I uttered to the reflection in the rearview mirror.

I replayed every detail—what we talked about, her unexpected chic appearance, and the gravity of this life event. Although, I understand Peggy's motives for confidentiality, it didn't make it any easier on me. Once again, another entire family of my blood relatives was completely unaware of my existence. I was the dirty little secret no one wanted to acknowledge. A sense of sadness lingered over me. A part of me wished that our time together had been more joyful and focused on our relationship, instead of the anger she harbored towards Father O'Reilly. If this was my first impression of the woman who gave me birth, I wasn't overly impressed but I did have grace and empathy given her pain over the years.

> Once again, another entire family of my blood relatives was completely unaware of my existence. I was the dirty little secret no one wanted to acknowledge.

On the other hand, I felt a heightened sense of relief at finally meeting my birth mother. It helped me fill in some of missing pieces of the puzzle regarding my identity. The empty cavity of unknown space was now full. I knew where I came from, why I was given up, and the purpose for the concealment.

> The search for my birth mother had led me full circle to recognize and appreciate the loving relationship I already had with my mom, Flo.

The force of realization overwhelmed me. I wanted my mom, my *true* mom, not the biological one. I wanted Flo, the woman who chose me and loved me and raised me as her own. The search for my birth mother had led me full circle to recognize and appreciate the loving relationship I already had with my mom, Flo. I was blessed more than I ever knew.

I felt like I was going to burst with all of this new information, and my foot pressed down a little harder on the pedal. I needed to get to a phone and share my experience. I couldn't wait to tell my wife Connie, my sister Karen, and my dear friend Krystal Joy. Throughout the day, I squeezed in talks with the girls during my breaks from the seminar. Their squeals and sighs mirrored my own excitement and we shared tears together at the culmination of all the years of searching. None of us could believe I had finally met my birth mother. I thanked them over and over, overwhelmed with emotion. I knew that without their support I could never have navigated this emotional roller coaster.

> I knew that without their support I could never have navigated this emotional roller coaster.

Chapter 16

KNOWING A GOOD MAN

A family is a risky venture, because the greater the love, the greater the loss... That's the trade-off. But I'll take it all.

Brad Pitt

A few days later, I got in touch with Father O'Reilly to describe my encounter and unsettling interaction with Peggy. I looked up his new number and jotted it down in my Rolodex. Then I dialed with trepidation. There was no private office number to call Father O'Reilly on anymore since he retired from Springfield, Massachusetts, down to Cocoa Beach, Florida, with his wife Florence. If she picked up I would be forced to come up with some lame alias because both she and his kids were still unaware of my existence.

Fortunately, Father O'Reilly picked up right away. After a round of pleasantries and catching up, I broached the real reason I had called.

"Father O'Reilly, I finally got a chance to sit down with Peggy and talk."

"Wow, Paul. How did it go? Tell me about it? How are you doing?"

I jumped in and recounted the details, knowing he was curious regarding Peggy's viewpoint and stance of past.

"Father O'Reilly, it's not pretty," I explained. "Peggy was, and is, indignant to this day. She carries an immense amount of guilt for getting romantically involved with a priest and giving birth to his son along with giving up her child for adoption. And—well, she pretty much blames you for virtually everything and feels like you got off scot free while she, in turn, was scarred for life."

Father O'Reilly hesitated in his response. I knew he was pondering what best to say in the awkward space. "Paul, I think it's great you finally got to meet your birth mother. That must have been a powerful moment and I'm sure it has answered many lingering questions for you. I can't say I'm not a little disappointed though, to hear that Peggy chooses not to accept any accountability for our indiscretion. But, I guess I can handle the blame for everything," he laughed quietly.

"Paul, don't ever forget the true blessing of our impropriety is you! The Bible says, '*Children are a gift from the Lord; they are a reward from him. Children born to a young man are like arrows in a warrior's hands. How joyful is the man whose quiver is full of them!*' Ps. 127:3-5 (NLT) God chooses children and you are no mistake!"

God chooses children and you are no mistake!

I deeply sighed and choked back the emotion in my throat. Father O'Reilly had an almost effortless way of reassuring me of both my value and worth. Maybe it was the priest in him, but I appreciated his perspective and the idea that God had me in mind all along.

We returned to the topic of my birth mother on many occasions and Father O'Reilly always answered my questions in a straightforward way. Over time, our relationship transpired and grew in strength. He became my mentor and one of my strongest advocates. After I moved from North Carolina to Scottsdale, Arizona, due to a corporate transfer, we continued to enjoy our long distance talks from Florida to Arizona on a monthly basis. I relished his insight and looked forward to our discussions. Father O'Reilly was always jubilant and in good spirits, and I believed he treasured our conversation as much as I did.

As the months and years flew by, he monitored the progress of my daughter and his granddaughter Amanda, checked in on my professional achievements, scolded me for traveling too much, and always kept an eye on my latest golf scores. We compared and competed over our latest cholesterol numbers and ribbed the loser—in this case the one with the highest number. It seemed like each and every call he would provide a detail or a crucial memory about my past or his past that helped fill in the blanks.

Unfortunately, Peggy didn't have the zest or inclination to build the type of connection my birth father and I cultivated over time. In fact, Peggy wasn't motivated in the least to keep in touch with me. I called her every six months or so to check in and make an attempt at a relationship, but her tone was frosty and guarded and I always felt like an intruder in her space. Her apprehension that her family would eventually find out about her past overwhelmed her and produced an overwrought and almost paranoid state each time we conversed. The phone calls were terse and one-sided and eventually it became very clear that I alone was driving any semblance of a relationship between the two of us. It finally came to a head one day. I was living in Scottsdale, Arizona at the time and called her to check in.

"Hi Peggy, its Paul, how are you getting on?" I paused, expecting a semi-cordial response.

But I was in for a surprise. In a firm and unyielding voice, Peggy informed me, "Look Paul, I'm sorry, but our association is not a good idea at this point in my life. I can't do these calls anymore. It's too emotionally draining on me. All the hurt and the pain from the past comes right back to the surface. Please refrain from contacting me again. I hope you understand and respect my wishes."

Stunned, I stammered out a weak "okay." Gathering my thoughts while pushing away the panic in my throat, I considered what to say in what might be our last conversation.

"I respect your decision and you won't hear from me again." My voice faltered, "I wish you the best."

The line went dead as Peggy disconnected. And just like that, our tentative relationship ended. I was baffled and angry. I could accept and even understand why she had given me up the first time around as an infant, but this was the final betrayal. How could a birth mother give up her child twice? She was blessed with a second chance—a sacred opportunity later in life to re-establish a relationship with her first and only son and then she chose, of her own selfish volition, to walk away once again. The only comfort I could find was in the Bible verse. "Can a woman forget her nursing child, and not have compassion on the son of her womb? Surely *they may forget*, Yet I (God) will not forget you. See, I have inscribed you on the palms of My hands." Isaiah 49:15-16 (NKJV)

If I had been looking for validation from my birth mother, this incident could have emotionally taken me out. Fortunately, because of my confidence in the love and security of my mom Flo, Peggy's behavior only highlighted the extreme difference between the two women. It confirmed that my adoption was not a mistake, but a divine intervention and God delivered me into the right woman's hands.

On the bright side, as my relationship with Peggy drew to a close, my bond with Father O'Reilly continued to grow and develop. I also drew closer to Father O'Reilly's sister, my Aunt Colleen. She was a rock for me through the years. She never wavered in her support and was instrumental in convincing Father O'Reilly to come clean and fess up that he was indeed my birth father. Before, during, and after the search, her encouragement allowed me to plow forward and seek the answers to my past and uncover my true identity.

One day, when I dialed up Father O'Reilly, his voice cracked repeatedly during the conversation and sounded strained. I poked at him for shouting too much—knowing this was the polar opposite of his gentle demeanor. Father O'Reilly hoarsely chuckled and brushed it off as a bad case of laryngitis. A few weeks later, his voice deteriorated even more. Now I was a bit more concerned, but Father O'Reilly assured me in a raspy but firm voice that his physician was running a series of tests and scans to get to the bottom of his chronic sore throat.

I was anxious to hear the results of his scans. Unfortunately the next call, confirmed my worst fears. Father O'Reilly was diagnosed with throat cancer and was quickly losing the ability to speak. Through strained and inflamed vocal chords, Father O'Reilly's croaky whisper sounded terrible and I could barely comprehend anything. I made sure to tell him how much I appreciated him and loved him during the call—not sure if it would be the last.

My position as the hidden illegitimate child put me in a precarious situation. I couldn't contact his wife Florence and the kids for status updates because they didn't know I existed. I also knew that now was not the time to let them know. They were under enough stress.

My position—as the hidden illegitimate child—put me in a precarious position.

My only option was to stay in close contact with my Aunt Colleen so she could keep me informed of her brother's treatment.

I was aggravated to put it mildly. I couldn't visit Father O'Reilly or talk with him because of his condition, and the sicker he got the more helpless I felt. I was shut out of my birth father's declining health and cancer battle. The deception made me sick to my stomach. I loved my birth dad and although I didn't agree with the cover-up, I respected him too much to divulge his secret.

Months slipped by. I prayed and hoped for a miracle but knew in my heart I would never see Father O'Reilly again. It tore me up.

One morning, I received a teary call from Aunt Colleen. "Hi Paul, I'm sorry I don't have good news. Sweetheart, Father O'Reilly passed away last night."

Although I knew it was inevitable, the finality of her words crashed into my heart. "Oh, okay," I garbled out weakly.

"Paul, the funeral service and burial will be at Arlington National Cemetery. It's a pretty spectacular place. I don't know if you remember, but that's where your birth dad was a chaplain for several years."

I vaguely recalled Father O'Reilly working there when I was kid, "Umm-hmm."

"I really think you should come to the funeral. You are his son. You have every right to attend and I'll be with you."

I knew I had to go. I wanted to pay my respect to the man who brought me into this world and cared for me from a distance for many years.

"Aunt Colleen," I sighed, "I really want to go, but won't my being there stir up suspicion? The last thing I want to do is cause some big controversy and be perceived as disrespectful to the family."

"You just stick with me Paul and we'll navigate this without hurting anyone, okay?"

I agreed and we worked on the details as I bit back my apprehension. The next few days were a blur of wrapping up work, planning for the trip and dealing with spontaneous emotional breakdowns. The tears were a shock and the sadness somewhat unexpected. I didn't foresee how much Father O'Reilly's death would affect me. The man wasn't even a significant part of my life until I was thirty-five years old and then only on a limited basis. However I tried to dismiss our relationship as minor- it wasn't.

However I tried to dismiss our relationship as minor— it wasn't.

There was an unexplainable bond between us from the first moment we connected at Friendly's. We looked alike, had similar mannerisms, the same smile, sense of humor, demeanor, and passion for life. Father O'Reilly had an incredibly amiable spirit and was so easy to talk to. It felt like I had known him my whole life. His passing, to my surprise, left a gaping void in my life.

Despite the fact that I deeply cared for my adopted dad and appreciated all he had done for me, we never developed closeness. Who knows? Maybe on some level he subconsciously resented that I wasn't biologically his child. Every time we got together a big painful silence ensued. We tried to find common ground to link us together, but truthfully, besides my mom there wasn't much! We were just too different, possibly too resistant, and the gap between us was too big to overcome.

Funny enough, I didn't know how much I truly longed for a loving male father figure until Father Ted stepped into that role. The ache I unknowingly carried around my entire life gradually healed over the years with Father Ted in my life. Even though it was only ten years, I'm thankful for the restoration and the impact of a loving birth father. If I hadn't taken the initiative to search for my birth parents I would have missed out on one of the most pivotal relationships of my life.

> If I hadn't taken the initiative to search for my birth parents I would have missed out on one of the most pivotal relationships of my life.

With the loss of my birth dad, I figured I had nothing to lose by trying to contact my birth mother one more time and inform her of Father O'Reilly's passing. Although I promised not to contact her, this seemed like an opportunity to give her valuable information and possibly reconnect. Ever hopeful, I hoped maybe her heart had softened in the interim years of our disconnection.

Peggy picked up on the second ring and I breathed a sigh of relief at avoiding my half siblings or her husband who remained in the dark.

"Hi Peggy, this is Paul. I'm sorry to bother you but I thought you should know that Father O'Reilly passed from throat cancer this last week. Were you aware?"

Peggy retorted with a frosty, "Yes!"

Her sharp exhale was stern and without remorse. "It serves him right for all the pain and suffering he has put me through all these years. I hope you don't expect me to be sad about Father O'Reilly's death because I'm not. In fact, I'm relieved he is finally gone!"

I had no words. How does one respond to such venom? I fumbled with the phone and finally just dropped it back onto the cradle.

Once again Peggy's true character revealed itself. Did it ever occur to this woman that I had just lost my birth father? Was her empathy so lacking she couldn't even acknowledge my sadness? And why, oh why, did I keep hoping that every interaction with her would be different than the time before? The truth was my biological mother was a very hurt and damaged woman, and hurt people tend to hurt people.

I wasn't too excited about making the next phone call either. I needed to call my adoptive parents and inform them of Father O'Reilly's passing and let them know I was flying out to the East Coast to attend his funeral at Arlington National Cemetery with

Aunt Colleen. The line went silent while they processed my words. All I heard was a sharp intake of breath. I tried to have compassion with their response or better yet, lack of response, but it was hard because I needed something from them I couldn't articulate at the time. My unmet expectations mingled with their conflicted emotions, a desire for acceptance (pitted against) their hurt and betrayal.

I'm not sure what happened to my dad after I shared the news. Maybe he dropped the phone and left the room? All I know is his piercing silence did little to soothe my already frayed nerves. I heard muffled communication with my mom. I know Father O'Reilly wasn't his favorite person now, but a little sympathy from my dad would have gone a long way.

My mom, on the other hand, took a few moments to gather her thoughts, and then got back on the line. I knew her kind words about Father Ted were forced, but I appreciated it all the more because it showed her depth of compassion for me. She even offered to go to the funeral if I needed her for support. I knew what a difficult position my parents were in. The whole situation was emotionally difficult to maneuver, and messy. Betrayals of this magnitude aren't easy to let go.

Unfortunately, a few days later my mom changed her stance. She called and pleaded with me not to attend. Maybe my dad got to her or fear overwhelmed her, but my mom was deeply concerned I might distract Father O'Reilly's immediate family from grieving if they were to discover who I was. She was pretty adamant I shouldn't attend out of respect for the family. And I must confess, I was pretty adamant right back.

I swiftly but respectfully reminded her that I was also considered an immediate family member as Father O'Reilly's firstborn son. I explained gently, that while I appreciated her concern, I was quite capable of handling this on my own. I reminded her Aunt Colleen would be by my side if I needed assistance. I dearly love my mom and she has always looked out for my best interests. However, in this case I didn't feel that she could possibly relate to what I was going through. She was being more protective of Father O'Reilly's family than of my need to be at the funeral.

Chapter 17
POMP AND CIRCUMSTANCE

Some people think that the truth can be hidden with a little cover-up and decoration. But as time goes by, what is true is revealed, and what is fake fades away.

Ismail Haniyeh, *The 51 Day War: Ruin and Resistance in Gaza by Max Blumenthal*

Aunt Colleen decided to introduce me as her godson at the funeral, with the guise of my attendance as a show of respect for the passing of her beloved brother. I thought it was a good plan because it was based partially on truth. I didn't have to conceal my identity; I simply didn't reveal the extent of my connection to Father O'Reilly. From my perspective, it seemed to me that my presence at the funeral would be the hardest on me anyway, not Father O'Reilly's family. They didn't know I even existed, but I, on the other hand, would be encountering my half brothers and half sister for the first time. And I couldn't react in any way—no small feat! Yes, I guess there was a slight chance someone might notice my resemblance to Father O'Reilly and put it all together, but more likely, in their current state of grief, they wouldn't even notice me. I would suffer through non-acknowledgement as Father O'Reilly's son and watch other family members openly share their emotions, while struggling to conceal mine. I felt a bit cheated—even in death I was the outsider.

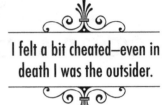

I felt a bit cheated—even in death I was the outsider.

I arranged to fly back to Washington, D.C., the night before the funeral and reserved a room in the same hotel as Aunt Chris and the rest of Father O'Reilly's family, including his two other sisters, Nancy and Mary, along with their husbands and kids. The entire family had driven down together, caravan style, from Massachusetts to Washington, D.C. As far as I knew, none of the other family members were aware that I was Father O'Reilly's son except for Aunt Colleen. I was looking forward to meeting my extended biological family, although my apprehension at discovery made me feel quite uneasy.

That night I joined the adults for a quiet dinner while the kids and grandkids went off and did their own thing. It was thoroughly enjoyable seeing everyone in the flesh after all of these years of getting glimpses of my birth father's life. I caught my breath a few times when I noticed striking similarities between Father O'Reilly's sisters and their strong resemblance to both him and me. Strangely I felt a sense of belonging I hadn't anticipated, and I even noticed eyes lingering on me with smiles a few times.

About halfway through dinner Aunt Colleen leaned over to me and whispered in my ear, "Paul, they know Father O'Reilly was your birth father."

I gulped down my food with a large swish of wine, trying not to fall off the chair.

Aunt Colleen grinned at me, "On the drive down from Worcester to Washington, D.C., I sort of let it slip. I guess the family secret has been divulged. Maybe when we get back to the hotel you can share with them the full story from your perspective."

Now I knew why all the eyes were on me during dinner. They were checking me out too for the resemblance I shared with their brother, Father O'Reilly.

I felt raw and a little vulnerable with the truth exposed, but since no one was rejecting me I leaned into the awkward moment. It took me a moment to raise my eyes, but when I did, smiles and even tears gazed back at me. It was comforting to know they cared. It dawned on me in that moment, how curious I was to discover what they knew and I didn't know about the past. This was a golden opportunity to uncover significant insight from Father O'Reilly's sisters about what happened from the very beginning.

We finished dinner and ambled back across the street to the hotel where we were all residing. Aunt Colleen spontaneously invited everyone to her and her husband Ray's room so I could share my journey with them more discreetly. Glancing around the room I was overcome with emotion—these faces that resembled mine, the

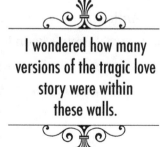

I wondered how many versions of the tragic love story were within these walls.

cerulean blue eyes and the silver streaked heads of thick hair—they were my blood-related family members. My story that began forty-four years ago included this cast of characters. They were all there from the very beginning!

My words were at first cautious. I wasn't sure of where or how to begin. Was Aunt Colleen going to moderate this scandalous topic and then I would jump in with my version of the story or did she want me to just go for it? After several minutes of painful small talk I realized Aunt Colleen was simply waiting for me to feel comfortable enough to open up. If anything, they were all waiting for me to make the first move and broach the subject that was consuming all their thoughts.

I knew I needed to tread lightly and be extremely careful not to offend anybody because Father O'Reilly was an adored brother, despite his moral lapse. Finally, with a small prayer for courage, I spoke up, "Aunt Colleen informed me over dinner that all of you are aware of the fact that Father O'Reilly is my birth father."

Everyone nodded and leaned in.

Bolstered by their encouragement, I went on, "I imagine you are interested in hearing about my journey to find my birth parents. Well, it started out for medical reasons…"

Smiles met my every word and I relaxed and persisted on. My biological family's keen interest made me decide not to give them the watered down Readers Digest version, but to dig in deep and thoroughly reveal the entire story, as I knew it.

As the story poured out of me, I sensed even more understanding and empathy than I had ever anticipated. If anything, they seemed sad for me because of the situation I was forced to endure over the years. Maybe, I was so tentative and cautious because of the negative reactions from Peggy and even my own adoptive parents. Certainly, I had been burned, but not here and not in this moment. Today I was met with love and acceptance. In fact, they were all on the edge of their chairs, anxious to hear my side of the story and all the details that went along with it.

Today, I was met with love and acceptance.

"I want to make it clear to you that from the very beginning the sole purpose of my search was to gain clarity regarding any medical issues that ran in the family. Specifically regarding elevated cholesterol, heart disease, and juvenile rheumatoid arthritis, because my daughter, Amanda had a bout with juvenile rheumatoid arthritis during my three-year search. I also want you to know I understand the volatile nature of what I've uncovered and the impact this scandal could have on the family. I certainly wasn't looking to uncover any big family secrets; I just needed to know my medical history. When I started this search I never expected to find out what I did."

Over the next two hours, I steered them through the aspects of my journey. Gasps erupted at times, many tears, and a few startling interruptions.

Mary, Father O'Reilly's oldest sister interjected at one point, "Wow, this is difficult to say, but I need you to know that Father O'Reilly actually came to me when Peggy was pregnant and asked if I would be willing to take the baby—you—into my family for adoption. I believe he hoped he could keep you in the family." Mary's voice cracked, "Paul, I...Stan and I just couldn't make it work at the time. We already had several little kids, and..." Mary's face crumpled in tears, "I'm so sorry for not taking you in. I feel like we made a huge mistake. I just didn't know how I would explain it to my parents. They would have suspected something fishy."

Then Father O'Reilly's other sister, Nancy, interrupted Mary's apology with one of her own. "Oh geez, Paul, I'm sorry, but Father O'Reilly came to us too and asked the same thing. He was determined to find a good and loving home where he could keep an eye on you. We told him no for the exact same reason. And Jack too (Father O'Reilly's brother who passed a few years before), I know Father O'Reilly asked Jack to take you in."

Aunt Colleen chimed in, "Well, he didn't ask me. Apparently, I was the only one in the dark here!"

Mary and Nancy gave a weak laugh.

Nancy nudged me, "You know Colleen was just a teenager then, right? Father O'Reilly wasn't going to destroy his baby sister's adoration. He asked me specifically not to tell Colleen or our parents."

Mary nodded in agreement, "Me too."

I could have talked all night, but I wanted to respect everyone's need for rest and it was almost midnight. The funeral was only hours away. I thanked everyone and suggested we wrap it up.

My aunts and uncles all embraced me with big hugs and expressed gratitude for sharing my personal story with them. They were all quite moved by the engaging story that intertwined us all.

Aunt Colleen gave me a big hug and whispered in my ear as I headed for the door, "I'm proud of you Paul. You know, I think they all knew about Father O'Reilly's predicament from the get go, but I'm sure you enlightened them on quite a few of the details. It filled in a lot of the holes." Aunt Colleen winked at me and patted me on the back.

"I can't believe we are burying my big brother in the morning." She murmured shaking her head in dismay. We shared a regretful look and I walked back to my room to dwell on all that had transpired.

Once I got ready for bed and hit the sheets, questions bounced through my head like a ping-pong ball.

"Oh great," I thought to myself, "I am never going to fall asleep."

Then I remembered the three-hour time difference from the East Coast to Scottsdale, Arizona and reached for the phone with delight. I dialed home and Connie picked right up. My wife and daughter Amanda were all ears about daddy's trip and wanted each and every detail about my evening with my "other" family.

Once I relayed all the news and said goodnight, I rested my head on the pillow again and tried to settle down, but sleep eluded me. I mulled over the events of the evening, mentally dissecting them one by one.

It had been a relief to finally tell my story to Father O'Reilly's sisters and their husbands, and I felt fortunate they chose to accept me, despite their guilt swirling around the past. They deserved to know the whole truth and I couldn't help but think how differently my life might have turned out if Mary, Nancy or Jack adopted me and raised me within the family. It would have been different, but I'm not sure if it would have been any better. My adopted family did a tremendous job loving and supporting me through the years and my regrets on that front were slim. It's probably the reason I hadn't searched for my biological parents earlier. It wasn't a pressing issue because I felt so loved.

I'm still uncertain why nobody informed Aunt Colleen after all these years. She was truly in the dark until I confided to her in the middle of my search. Her sisters and brothers sure took their vows of secrecy seriously. Clearly, they protected their brother—even from his little sister. I guess some secrets become so familiar that we just don't think of revealing them.

I woke up bleary eyed and groggy the next morning, but it didn't take long for the adrenaline to kick in and pump me full of anxious energy. I carefully dressed in a conservative black pinstripe suit paired with a starched white shirt and crimson red tie. I stared in the mirror, and for once, hoped I didn't look too much like my birth father. I then headed downstairs to join the family for breakfast before the funeral service.

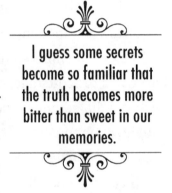

I guess some secrets become so familiar that the truth becomes more bitter than sweet in our memories.

I spotted Aunt Colleen seated with her husband Ray and the whole crew from dinner the night before. They were boisterous and loud as large families always are. I envied their closeness before it dawned on me that I was now semi-included as Father O'Reilly's son—their nephew. The thought gave me goosebumps. I gave a weak wave, found a chair and pulled it up to the table. Despite the warm reception, I still felt a little raw and vulnerable from spilling my guts only a few hours earlier, but their cheerful smiles quickly soothed my fears.

One by one, in hushed voices and whispers my aunts and uncles reiterated how much they appreciate me sharing my story with them the previous night. I got the picture that my identity was not yet public knowledge to the whole table—namely their children—so I treaded lightly on my response. Their kids, my cousins, most of them in their twenties were busy catching up with one another and didn't take too much notice of the stranger with their aunt. Although I recognized a few from dinner the night before, there were several I had never been introduced to.

"Kids, this is Paul," my Aunt Nancy asserted, giving me a subtle look with a tiny wink. "He is your Aunt Colleen's godson."

My Aunt Nancy introduced her children to me. I shook their hands while in a daze, "Nice to meet you."

It felt surreal, like I was watching a play unfold from a balcony seat. Here I was, forty-four-years-old meeting my cousins for the first time. Except they didn't know I was their cousin. The tension of my identity and continuing to hide it made me feel a bit like a fraud, but I knew it was a role I needed to play the entire day if I was to get through it without upsetting anyone.

After a large breakfast and shared recollections of Father O'Reilly, we all headed out to the cars and then caravanned across the street through the stately entrance to Arlington National Cemetery. I took in the majestic arches and imposing grounds while fighting back the tears. The magnitude of the moment snuck up on me. I had been so busy with all the preparation and the angst of the family meetings, I hadn't thought much about the sadness of losing my birth father. How would I hold it together when I met my brothers and sister for the first time? Could I really pull this off? I rubbed at my cuff and pulled at my collar as if loosening my tie might somehow dispel the apprehension and sadness welling up in my heart.

I had no clue what the protocol was for a funeral of this scale, so I simply followed my Aunt Colleen and Uncle Ray into the picturesque chapel at Arlington National Cemetery. Despite the crowd of people lining up, the sanctuary was rather still. Light streamed in through the ornate stained glass windows and bounced off the rich mahogany pews. Elaborate flower arrangements crowded around the ivory altar and the scent of jasmine and roses mingled with the musty aroma of the old church. We sat about five rows back from the first pew on the right side of the aisle where the immediate family was placed. Within a few short minutes, the rows filled up to capacity with family and friends, and with many more cramming in the back.

It could have been the intense emotional strain, but I had a sense that I was being watched. I took a deep breath, shook off the paranoia, and tried to remain detached while choking back my real feelings. This wasn't the time for speculation. I wouldn't allow my emotions to betray me during the service.

Rising out of the stillness and the quiet chatter around me, the first peals of music reached my ears.

It's beginning, I thought to myself. Breathe, Paul. Breathe.

I exhaled deeply and looked behind me to find the source of the melody. I spotted the organist playing from the second floor balcony, in the rear of the church, and then a flash of something vivid red caught my eye and I gasped.

Directly to my left, my birth father's casket draped in an American flag was being rolled down the aisle toward the altar by full dressed Army Corp of Engineers officiates. Father O'Reilly's wife, Florence and their three children—my two half brothers and half sister—sedately walked into the sanctuary from the side entrance and took a seat in the front row.

I struggled to maintain composure and not get too choked up in public, but my sentiment was overwrought and thin. There was a part of me grieving not only for my birth dad, but for the loss of my place in the family.

> There was a part of me grieving not only for my birth dad, but for the loss of my position in the family.

The pomp and circumstance of the event blew me away. My Aunt Colleen leaned over and explained, "Paul, Father O'Reilly's years in the army and service as a chaplain here at Arlington rewarded him a full military honors funeral service, similar to what a U.S. president receives upon his passing."

In that moment, I think I understood why my mom was so concerned about my attending the funeral. Father O'Reilly was sort of a big deal, a celebrity even, and everyone, including my mom, still went out of their way to protect him. The program applauded his bravery and selflessness in the midst of hellish conditions on the front lines. Father O'Reilly was not only a beloved priest, he was a national hero. And no one, including my mom, wanted to take down a good man who made a few bad decisions, but paid for them dearly on a battlefield.

After the service Father O'Reilly's casket, adorned in the American flag, was placed on a horse drawn carriage pulled by eight Clydesdale horses to the burial site. His funeral was the most honorable type of service performed at Arlington National Cemetery. It is reserved for presidents and heads of state, and in this case, for a war hero who was being honored for his service as chaplain at Arlington National Cemetery.

The immediate family was invited to walk behind the carriage while the remainders of the attendees were invited to stop by the Officers Club at Arlington National Cemetery for a post-service reception.

I wasn't sure what to do but Aunt Colleen squeezed my hand and motioned for me to follow her. It was an unforgettable moment. I couldn't believe I was included in the family. In a small way I got to be there for my birth dad. Once we arrived at the burial site, I discovered I inadvertently stood right next to Father O'Reilly's wife and my half siblings. I tried to remain incognito but playing it casual at a gravesite is not so easy. I avoided eye contact and looked anywhere but directly at them. I wondered if they were as curious about me as I was about them. Did they wonder, who is this strange guy? And why is he with the immediate family? But as awkward and uncomfortable as it was, I still relished the scene. Moments after the entire family assembled, the chaplain who performed the service began to pray. I closed my eyes, bowed my head, and solemnly recited the Lord's Prayer. I had yet to look up when the first shot rang out of a 21-gun salute, a soldier was playing poignant taps on his trumpet, and each shot ricocheted through my head and my heart. The Army Corp lined up in front of us in solemn respect holding their hands up to their forehead in a salute. This was the point that I felt my grief overcoming me and I battled to not lose it in front of the entire group. I did everything I could think of to hold it together.

Fortunately, the years of business training came back to me. Focus on something, I thought. I looked out into the distance, my eyes roaming for anything of substance to settle on, and to my surprise the tip of the Washington Monument was peeking above the trees of Arlington directly behind Father O'Reilly's gravesite. I'll never forget that view—it represented a calm sea when so much sorrow and conflict stormed within me.

I whispered my own personal prayer and farewell to Father O'Reilly. By God's grace, my demeanor remained stoic as they lowered him into the ground. I truly don't know how I held it all in but the fear of causing a scene and disrespecting my birth dad's memory was greater than the raw hurt pressing tight against my chest. I looked around as friends and family gathered up their belongings,

How ironic? —I was so close to them, but a million miles away all at the same time.

embraced, and swiped away teary eyes. Their encouragements and endearments went out to the family standing next to me. I could have reached out and touched them if I chose to. How ironic! I was so close to them, but a million miles away all at the same time.

Over and over I heard people affirm Father O'Reilly to his sons and daughter in respect for the great man he had been. And each time they patted Patrick O'Reilly Jr. or Bill, I felt the loss of my place in the family. I yearned for the acknowledgement I would never receive as a son of a great man.

Aunt Colleen ambled up then and took hold of my arm, guiding me to the car intuitively knowing I was struggling. On the way over to the reception at the Officers Club, she quietly shared with me more about Arlington and Father Ted's role as a chaplain after the war. Then she pointed out the quaint house Father O'Reilly lived in on the perimeter of the grounds. I took in the picturesque tranquility and breathed a sigh of thankfulness. I was glad his final resting place was Arlington, his former home—although I knew Father O'Reilly would tell me it was only a temporary place compared to his eternal home. I knew Father O'Reilly had made peace with God and this comforted me. At that moment, it felt like I could almost hear his voice in my mind speaking the Scripture he would often quote, "*Most assuredly, I say to you, he who believes in Me (Jesus Christ) has everlasting life.*" John 6:47 (NKJV)

The banquet room at the Officers Club was typically reserved for high-ranking officers from the U.S. Army, Navy, Air Force and Marines. As we walked upstairs there were austere portraits of previous US Presidents, generals and other men who lost their lives serving this country. It felt like the White House of cemeteries.

I sat down at a table with Aunt Colleen, her sisters and their husbands directly next to the immediate family table, including Father O'Reilly's wife, their three children and spouses. All of Colleen's, Nancy's, and Mary's kids, my cousins, were seated at a table on the other side of us. I pinched myself from excitement. This was too much! I was in a wonderland of secret family espionage; my incognito state afforded me the privilege to watch my own family relatively unnoticed. Trying not to be too obvious, I glanced through the corner of my eye at my cousins and half siblings. I couldn't help but be fixated, even mesmerized, by the nose of a cousin or the eyebrows of an uncle. I strained to hear how they laughed, studied their personalities for similarities and mannerisms, and even watched how my brother held his knife and fork. I felt like Marty in the movie, Back *to The Future* when he meets his teenage mother and relatives for the first time. In my family's eyes I was a stranger, but to me the moment was surreal and I relished in it.

Once the banquet room reached capacity and the crowd of family and close friends settled in at their tables, Father O'Reilly's wife Florence and their three children, Patrick Jr., Bill and Mary stood up and thanked everyone for joining them in celebration of Father O'Reilly's life. Then they made their way around to each table to personally thank each person for their support and willingness to come all the way to Washington, D.C., to attend Father O'Reilly's service. Almost everyone in attendance lived some distance away, so each guest had sacrificed to be here. The idea that my birth dad influenced so many lives in a meaningful way made a profound impact on me that day. His life of dedication and service lived on in the people he impacted spiritually.

> The idea that my birth dad influenced so many lives in a meaningful way made a profound impact on me that day. His life of dedication and service lived on in the people he impacted spiritually.

Before I knew it, I was staring into the face of Florence. She was an attractive older woman who looked adrift in the role of hostess to a party she never wanted to attend. Her eyes appeared glum but her smile was genuine. When she reached for my hand to shake, my legs started to tremble and I almost hyperventilated. I took a mouthful of air and feigned a reserved

> The harsh truth was they didn't know I existed, so they weren't looking to make any connections to their dad.

but congenial disposition as Aunt Colleen introduced me as her godson to each and every one of my siblings individually. We exchanged handshakes and cordially smiled at one another as I offered my condolences. I'm not sure what I expected from them but their pleasant, and yet, detached receptions didn't make my experience any easier. The harsh truth was they didn't know I existed, so they weren't looking to make any connections to their dad. They didn't have a big aha moment, or recognize my true identity and rush into my arms claiming their long lost brother who had resurfaced from the past. I was just another face on a terrible day, a day they had wished would never come.

The idea that my birth dad influenced so many lives in a meaningful way made a profound impact on me that day. His life of dedication and service lived on in the people he impacted spiritually.

The pressure of the moment boiled under the surface. Despite the positive self-talk I peppered myself with all day long to keep from losing it, I simply couldn't handle another minute. I politely excused myself for a breath of fresh air and headed with haste for the door. The cool air soothed my frayed nerves and I turned up my collar and headed down the road to walk off my irritation. The absurdity of the circumstances aggravated me to no end—here I was in front of my half brothers and half sister for the first time in my life and I had to withhold my identity from them. Glancing down at the

The harsh truth was they didn't know I existed, so they weren't looking to make any connections to their dad.

fork in the road, I noticed the sign for the parsonage and veered towards it. This was Father O'Reilly's old residence, the one Aunt Colleen pointed out during the ride over from the burial site. He had lived there during the years he was chaplain at Arlington National Cemetery. Maybe a moment of recollection on Father O'Reilly would refocus me.

They didn't have a big aha moment, or recognize my true identity and rush into my arms claiming their long lost brother who had resurfaced from the past. I was just another face on a terrible day.

As I turned the corner, gravel crunching under my quick steps, the building came back into view, a lovely two-story traditional home, refurbished over the years and immaculately maintained along with the beautiful grounds of Arlington National Cemetery. I tried to envision what life was like for Father O'Reilly during that season of his life and ministry. It was after his tenure in the Vietnam War and before he left the priesthood to marry Florence and have children, I wondered where I was during those years? Did he miss me? Did he feel like he missed out on fatherhood and long for a different life? I wondered if he ever thought of me when he read the Scripture, *"Children are a gift from the LORD; they are a reward from Him."* Ps. 127:3 (NLT)

The absurdity of the circumstances aggravated me to no end—here I was in front of my half-brothers and half sister for the first time in my life and I had to withhold my identity from them.

I wandered around the building until my head cleared and then started to make my way back to the Officers Club, knowing

Chris was probably concerned about my whereabouts. I arrived just in time to say my goodbyes and dole out hugs. We set off for the hotel and returned to our individual hotel rooms to rest and recuperate after a draining and emotionally taxing day.

I took off my coat, tossed my shoes on the floor and climbed into the bed completely spent. An hour or two later, the phone rang and I picked it up.

"Hello—this is Paul," I said.

"Hi Paul, its Patti, Aunt Mary's oldest daughter. I wanted to know if you were up for joining me and all my cousins for dinner. We are heading into Georgetown. If you want to come we are meeting in the lobby downstairs in about an hour."

I stuttered a bit, excited for the invitation, "Uh, yes! I would be glad to join you. See you soon."

I sat up and rubbed my eyes. This would be another interesting adventure—hanging out with more family who didn't know who I was.

The crew of cousins, including me as the covert cousin, assembled in the lobby and then took off for a brisk trek to the train station where we boarded a train to nearby Georgetown. We stumbled upon a trendy but casual restaurant in Georgetown that wasn't overcrowded and could seat all fifteen of us in a reasonable timeframe.

As the new guy in the group of cousins which were accompanied by spouses or significant others, most of my time was simply spent introducing myself and chuckling at inside jokes and stories that were being retold on my behalf. I reveled in the closeness and familiarity of a big rowdy Irish family. Everyone was affable and strangely enough, from my perspective, truly bent on getting to know me. I was more than a little flattered and blown away by their warmth. I thought I might feel like an intruder, but if anything I was the object of their attention for some unknown reason.

Dinner passed all too quickly and the conversation flowed along with the wine. My cheeks were flushed from laughing when all of a sudden the table quieted. I craned my neck and pulled my chair in for a better view. It seemed that everyone's eyes were on Patti—the most senior cousin.

"So—Paul," Patti inquired, "what did you think of our uncle, Father O'Reilly's service today?"

I took a sip of water and pondered how to respond. It felt like all eyes were on me waiting in anticipation for my answer. (No pressure!)

"Well," (I was treading lightly), "I was extremely impressed by the pomp and circumstance. I certainly wasn't expecting anything so formal. The only time I've seen a horse drawn carriage pulling a casket, a 21-gun salute, and taps playing was at the funeral when John F. Kennedy was buried at Arlington."

I nodded my head. "Wow, well, it was beautiful and I'm honored to be here with you guys," I looked around the table and grinned sheepishly while opening up my arms in a gesture to include them all. "It's also really cool to hang out with a big family like yours. I love it! It's wonderful how you care and support one another. I'm not used to anything like this coming from a small family with one younger sister and only a few cousins."

"Now Paul, that's not completely accurate, is it?" Patti said with a wink.

I didn't dare reply. Was she going to make some big revelation about me at the dinner table? I glanced around the table. All eyes were locked on Patti.

"Ummm, Patti, what do you mean?" I weakly threw out.

"Paul, we all know better, we know that you are originally from a much larger family—aren't you?"

It took a moment for me to digest her words. Did this mean that all of the fifteen cousins sitting around the table knew I was Father O'Reilly's first son and their long lost cousin?

"You've got to be kidding me! You all know my secret story? How long have you known?"

Patti jumped in, "My mom told us awhile back," she motioned to her brothers and sisters. "But she swore us to secrecy given the family sensitivity and possible scandal."

Patti gave me a sly expression, "I think Uncle Denny (Nancy's husband) gave away your secret this morning. Apparently, Uncle Denny gave an abbreviated version of your story, from the night before, to his boys and they told everyone else in about two minutes. Before breakfast ended I'm pretty sure everyone knew."

I was flabbergasted. "WOW!" I shook my head and beamed. I guess that explained why I felt so exposed at the funeral. People were actually staring at me and it wasn't my imagination."

One by one, each of my cousins got up from the table and embraced me—most with tears streaming down their faces. At this point, even through my own joyful tears, I felt emotionally overwhelmed. With the last hug, a toast and a cheer went up from around the table, "Welcome back to the family Paul!"

It was almost sunrise before I got back to my hotel room but my exhaustion was fueled with pure bliss! It was such a treat to spend precious time getting to know my cousins. Each and every one was fascinated with my birth story narrative and subsequent search.

I had phone numbers and addresses in my pocket and I looked forward to remaining in contact with them and further developing our relationships. Of course, I couldn't sleep again; I was way too riled up by contemplating everything that had transpired that day. I tossed and turned in my bed for a few hours until it was time to get up, get dressed, and rejoin the family for our last meal together. There were many warmhearted bear hugs and emotional goodbyes to everyone over breakfast before I headed to the airport and flew home to Scottsdale, Arizona. The plane ride felt like prolonged imprisonment because of the prized information I carried within me. I felt like I had a giant present waiting to be opened. I was ready to explode with the latest events of my own real life mini-drama. From beginning to end, there was so much to share with the people I loved about this mind-boggling trip to Washington, D.C.

I tried to calm down on the flight and rest after days of running on adrenaline, but thoughts replayed over and over in my head: the sadness of my birth father's death, the blessing of reuniting with my family and meeting aunts, uncles, and cousins who wanted to be in relationship with me, the pageantry of the funeral and the aggravation of remaining anonymous to my own siblings.

When I arrived home Connie and Amanda were eager to hear the details of my trip, as were my adoptive parents, and good friends Krystal and Reuben Joy. And needless to say, I was eager to disclose the tale.

Chapter 18

FAMILIAL OBSTACLES

COLONEL JESSEP: *You want answers?*
KAFFEE: *I think I'm entitled to.*
COLONEL JESSEP: *You want answers?*
KAFFEE: *I want the truth!*
COLONEL JESSEP: *You can't handle the truth*!

A Few Good Men, Columbia Pictures, 1992

As time went by, the pain of the initial loss of my birth father waned, but the desire to know my two half brothers and my half sister grew. I discussed with Aunt Colleen the possibility of contacting them. It was difficult to estimate when they might be the most receptive to contact. There aren't any manuals to rely on when it comes to sensitive matters of the undisclosed past. The last thing I wanted was to offend

There aren't any manuals to rely on when it comes to sensitive matters of the undisclosed past.

them or be perceived as disrespectful in the wake of their father's death. Aunt Colleen, always the wise mentor, suggested I wait at least two years to allow them ample time to heal. I figured after patiently enduring forty-four years without knowing them, I could certainly wait another two.

However, there was some sense of urgency because all of the cousins were in the know about the big cover-up. Both Aunt Colleen and I realized it was only a matter of time before the story leaked to Father O'Reilly's kids and his wife Florence. At least Florence and one of her sons resided in Florida, so no one was going to bump into her at the local coffee shop or grocery store and reveal the scandal. We decided to go with our initial instinct and wait it out.

But waiting it out wasn't as easy as I imagined; it was actually brutal. Logically, I knew it was the right thing to do under the circumstances, but emotionally once I saw their actual faces and shook their hands—well, I wanted more. I wanted the potential of a relationship with my siblings.

It was actually 2005 before we made the first contact. This was three years after Father O'Reilly's death. Aunt Colleen believed the most appropriate course of action was to gently break the news to Father O'Reilly's wife, Florence. And then she could communicate the information to her children and tell them about their older brother, whom they never knew existed, and who resided on the West Coast.

The yearlong delay was due to unanticipated obstacles; namely, Aunt Collen's struggling to reach Florence by telephone. Florence didn't initially respond to Colleen's overtures, so Colleen ended up composing a letter encouraging Florence to call her regarding a highly sensitive family issue they needed to discuss. Florence eventually called Colleen, but she wasn't very receptive to the news. In fact, she was incensed. Her heated response and overall avoidance of Colleen left us all uncertain if Florence was in the dark on this issue or not. Was it possible at the end of his life Father O'Reilly shared the truth with her? All we knew was the secret was out and it was a difficult secret for Florence to deal with.

Even though Florence was evasive with Colleen, once the details were disclosed Florence moved into action. She promptly called a family meeting at her home in Florida to share the shocking revelation with her children. Patrick Jr. flew in from Washington, D.C., Mary from Boston, and Bill who lived nearby, all arranged to convene. Florence also requested Colleen join them and fly in from Boston to help with the specifics that the children would want to know.

Once the family assembled, Florence asked Colleen to walk the kids through the entire story from beginning to end. Their response was anything but positive. They were infuriated. The boys were deeply offended at their dad for not sharing this with them while he was still alive and apparently livid at me for bringing this distressing news to the surface. According to Aunt Colleen, Mary was the most receptive to the idea of a brother—certainly disappointed in her dad, but also excited to hear she had an older brother out there somewhere. Mary requested to look at the photos of Amanda and me that Aunt Colleen had brought down to Florida with her. The boys, on the other hand, refused to even glance at the photos.

It wasn't the reaction I hoped for; actually it was probably the worst scenario I could imagine. For so many years I anticipated a reunion of sorts, anticipating they would want to know me, as I desired to know them. But their response was not in my control and all I could do was surrender to time and hope for a change of heart, and maybe a little curiosity on their part about me.

Aunt Colleen recommended I give them time to process the hurt and betrayal of Father O'Reilly before I try to reach out and contact them. I heeded her advice and waited several months before I wrote a letter to Florence introducing myself and explaining my sincere motivation of only wanting to introduce myself to my half brothers and half sister. Unfortunately, she responded to my letter with a prickly note expressing her dismay at my reaching out to her or her children. She also made it clear how disappointed she was that I attended Father O'Reilly's funeral service without her or her children's knowledge. Ouch!

Always my defender, Aunt Colleen stood up for me and explained to Florence that I had a right to attend the service since Father O'Reilly was my birth father. And as far as informing her family just prior to the funeral service, it would have been reprehensible under the conditions. Rather than agreeing with Florence's accusations, Colleen pointed out that actually I had been highly respectful towards all those concerned. Colleen reiterated that I simply tried to fly below the radar and remain anonymous while attending my birth father's funeral.

> For so many years I anticipated a reunion of sorts, anticipating they would want to know me, as I desired to know them. But their response was not in my control and all I could do was surrender to time and hope for a change of heart—and maybe a little curiosity on their part about me over the years.

I've thought about my decision over the years and reflected on my choice and truthfully, I have no regrets. There are no written rules or correct protocol as to what is right or wrong in an emotionally volatile situation like this one. I simply tried to follow my heart, exercise good judgment, and remain sensitive to Father O'Reilly's immediate family.

There are no written rules or correct protocol as to what is right or wrong in an emotionally volatile situation like this one.

Almost thirteen years later, on the anniversary of Father O'Reilly's passing, I still have had no contact with Patrick Jr., Bill or Mary. I hoped and prayed over the years that at least one of them would take the initiative to reach out to me, but I imagine they all have their own reasons for not doing so. Some wounds and betrayals are hard to let go. I know Aunt Colleen is still planting seeds of restoration on my behalf. Perhaps when my siblings get older or after their mother passes away they will feel more at ease getting in touch with me. I know if I were aware I had an older brother out there somewhere, I would not hesitate to reach out and introduce myself. Despite the obstacles to reconciliation, family is family and I believe it's worth pursuing.

I may just pick up the phone one day and call my half brothers and half sister individually. Maybe speaking to them in person and having them hear my voice might overcome some of their reluctance. This journey of discovery has taken me down some crazy roads; maybe this is the next one I'm to pursue. Perhaps they're simply waiting for me to call!

Epilogue
INTERVIEW WITH PAUL

2015—Paul Aubin is currently the Men's Pastor at Mariners Church in Irvine, California—a large mega church ministering to thousands in the Orange County area. He lives with his wife Margaret and they are parents to a blended family including 7 children and five grandchildren. Paul left the pharmaceutical industry in 2009 to pursue ministry leadership full-time. I sat down with Paul and asked him some follow up questions regarding his birth parents search and the impact it made on his life, vocation, and relationships.

SAMANTHA: Paul, how has your journey to find your birth parents changed you?

PAUL: The pursuit of my birth parents gave me clarity on my background, my family tree, and the events that prompted the adoption in the first place. There's always a story behind an adoption and I wanted to know mine. I yearned to know my nationality and identity. I believe every adopted child seeks the missing pieces of the puzzle to their back-story and identity. We all ask, "Who am I?" My search helped me answer these imperative questions.

This journey also increased my faith dramatically and opened up a profound spiritual realm in my life. This quest put me on a fast track towards God. There were too many mysterious twists of fate and strange coincidences along the way that I couldn't explain. God's hand was at work and evident in every aspect of my search. Eventually, I couldn't deny His presence any longer and I reached out to Him. This gave me a sense of completeness and filled an aching void in my life.

Lastly, I discovered a sense of determination I didn't even know existed within me. I had courage and a persistence to pursue the truth at all costs, even if the truth had to hurt some people including myself. Sometimes I felt like Sherlock Holmes or a character from a movie on a treasure hunt, relying on an inner strength of steel that had to have been from God. Honestly, I felt like giving up several times throughout this journey because it was murky territory. But my resolve and willpower persevered and I learned a great deal about myself along the way. I am grateful I stayed the course even though it wasn't always pretty.

SAMANTHA: Tell me about the spiritual aspect of your search. Your birth father was a priest and now you are a pastor. For over 25 years you worked in the corporate world of the pharmaceutical industry. That's a big leap! What caused the shift?

PAUL: This search put in motion a spiritual story for me. I had clear evidence, from that pivotal moment back in the Marriott hotel when I stumbled upon the adoption convention, that God was directing my steps. I was a lukewarm Catholic before this started. I grew up in the church attending mass every single Sunday along with all the celebrated Holy Days. However, it seemed more of a ritual and routine to me with very little spiritual connection. It

Epilogue

was a place to socialize with all my buddies from the neighborhood and friends of my parents but nothing more than that. It wasn't personal to me, I merely felt like I was checking a box. It wasn't until the summer of 1997, when my job transferred me to Scottsdale, Arizona, that my boss, Craig Fitzgerald, informed me of a church in the area called Scottsdale Bible Church he encouraged me to attend. This was the very first non-denominational church I had ever attended, and, wow, what a difference from what I had experienced as a young boy attending church in West Babylon, New York. I felt like the pastor was speaking directly to me the first day I attended, and he inspired me to keep coming back. I just felt a real connection to God like never before. A few months later there was an opportunity to accept Jesus Christ as my Lord and Savior by saying a prayer and being baptized. That was the beginning of my new spiritual journey, as I grew in my faith various leaders of the church started approaching me with opportunities to lead in the men's ministry. It took about six years before I felt worthy of their invitations and started giving men's ministry leadership serious thought. While I was considering a leadership position within the church, I accepted a new job in Southern California and in 2003 I entered pastoral ministry at Saddleback Church in Orange County, California. As my birth father said, "The oldest son is destined to be a priest (or a pastor in my case!)." Clearly, going through this journey where I felt God's direction and comfort had a huge impact on my faith. Even greater impact was the surprise discovery that my birth dad was a Catholic priest.

This discovery was almost like a spiritual weight or responsibility in my DNA. It was a healthy weight that inspired me to be a better man and take the focus off of my own needs and shifting them to be a man who serves others. It's my legacy and I felt accountable to pass my gift on. I believe I was born for a reason and there is redemption in my career by ministering to others. Would God do all He has done for me to let me idle away as a pharmaceutical executive and play golf the rest of my life? I don't think so! I now get my joy from witnessing other men going through the same life-changing transformation that I have. Marriages are becoming healthier, dads are becoming more engaged with their kids, and our community has more men of God.

> I would check your motives before you even begin the first name search on Google. Why are you doing this? Is it for selfish reasons? Because overturning secrets and scandals is a larger issue than just you. There are people's lives at stake and it's not a time for revenge or bitterness.

SAMANTHA: What would you recommend to someone searching for birth parents?

PAUL: First of all, keep your expectations realistic! Try to understand this is not all about you and be highly sensitive to all those involved. In fact, I would check your motives before you even begin the first name search on Google. Why are you doing this? Is it for selfish reasons? Uncovering secrets and potential scandals is about more than just you. There are others lives affected and it's not a time for revenge or bitterness.

Another factor to keep in mind is the *reasonable* outcome. As a child, I mentioned earlier in the book that I had outlandish ideas of who my biological parents were. Unfortunately, the cruel realities of life can be harsh and there is little room for childish

Epilogue

fantasies and imaginations. Babies are sometimes given up not because they are unwanted, but because of terrible obstacles to overcome in the raising of a child. Your birth could have been from rape or a drunken night of debauchery, maybe an affair or a teenage girl too young to care for her child. Your parents might not even be alive. Please, remain practical and don't expect your birth parents to be kings and queens or you will most likely be highly disappointed. For those of you who search and are disappointed, think about what the Bible promises you, *"Even if my father and mother abandon me, the LORD will hold me close."* Ps. 27:10 (NLT)

Babies are sometimes given up not because they are unwanted, but because of terrible obstacles to overcome in the raising of a child. Every chess piece you move impacts the board.

Also, keep in mind that if you do find your birth parents their response might be negative. It's highly possible you can spend a fortune and years searching for a birth parent only to face rejection. And on top of that, your adopted parents may be hurt that you even went looking for your birth parents. Every chess piece you move impacts the whole board.

Expect difficulty and surround yourself with people who will support you. I have to admit the journey of finding my birth parents took a heavy emotional toll on me. Nancy helped prepare me for potential conflict, but it was the people on the road with me that pointed me back in the right direction during the rabbit trails that I chased in futility. I learned to leave my expectations at the door—there were no guarantees of the outcome of this adventure. I discovered that real people and real relationships are messy and closed adoptions are, generally speaking, closed for a reason. Having anchored relationships with people in your life that will pick you back up and dust you off, pat you on the back, and give you an attaboy; it will make all the difference in the world. It did for me. I'm not sure I could have persevered through a three-year search without my support network.

> Having anchored relationships; those people in your life that will pick you back up and dust you off, pat you on the back, and give you an attaboy, make all the difference in the world

The good news is that if you uncover the truth you will have a more complete sense of identity. You will know who you are and where you came from. And there is always the potential for family relationships to form. I was blessed with an incredible relationship with my birth dad for ten years and a large extended family that cares about me! One of the wonderful benefits of my searching journey has been the discovery of an enormous extended family of cousins, aunts, and uncles that I never knew before my search. There are fifteen first cousins from Ted's sisters Mary, Nancy, Colleen, as well as his brother Jack. They each now have numerous children and grandchildren of their own. It's so comforting to know that I not only have extended family within my adoptive family, but my search has opened up relationships and connections with my birth father's family. As an adoptee I once felt abandoned, but now my wounds have healed and I feel like I belong. Unfortunately, I live clear across the country in California and I don't have many opportunities to see this huge family in New England, but it's comforting to know that they are there when I need to call upon them. It's my hope and prayer that someday I'll have a similar experience with my birth mother's children and extended family.

SAMANTHA: What was the hardest pill to swallow throughout the search?

PAUL: There were more than a few! I struggled with the theory that my adoptive parents were not forthcoming with all of the information I assumed they knew. I also took my birth mother's rejection pretty hard. Her refusal to acknowledge me not only ruined our tentative bond but also denied me the opportunity to meet my two half sisters without causing more chaos. To this day, I still have no response from any of them. Then the dismissal of my three siblings on my birth dad's side who know the truth and still choose not to have a relationship with me is a sore spot.

SAMANTHA: Okay, last question! Paul, why did you write this book?

PAUL: Everyone I told this story to encouraged me to share it. People are moved by my story. We all have secrets that come back to haunt us and I found myself in a story much bigger than myself. It truly is a gripping story. Furthermore, I felt a responsibility and a burden to share this journey with other adoptees. Maybe they can avoid some of the landmines I encountered and take a few tips away with them. I would also encourage anyone searching for their parents to consider finding a coach along the way. Relationships are complicated and I certainly wasn't prepared for the hurdles I was up against. An expert can help make the relational impediments easier to navigate. In your search when you run into difficulty remember, *"We can rejoice, too, when we run into problems and trials, for we know that they help us develop endurance. And endurance develops strength of character, and character strengthens our confident hope of salvation. And this hope will not lead to disappointment. For we know how dearly God loves us, because he has given us the Holy Spirit to fill our hearts with his love."* Rom. 5:3-5 (NLT)

CPSIA information can be obtained
at www.ICGtesting.com
Printed in the USA
LVHW111527250422
717158LV00014B/86